BLOCKCHAIN FOR GOOD
The Transformative Impacts on Industry, Community and the Planet

Editors
Shoufeng Cao
School of Design, Faculty of Creative Industries, Education & Social Justice
Queensland University of Technology, Brisbane, Australia

Marcus Foth
School of Design, Faculty of Creative Industries, Education & Social Justice
Queensland University of Technology, Brisbane, Australia

CRC Press
Taylor & Francis Group
Boca Raton London New York

CRC Press is an imprint of the
Taylor & Francis Group, an **informa** business
A SCIENCE PUBLISHERS BOOK

First edition published 2025
by CRC Press
2385 NW Executive Center Drive, Suite 320, Boca Raton FL 33431

and by CRC Press
4 Park Square, Milton Park, Abingdon, Oxon, OX14 4RN

CRC Press is an imprint of Taylor & Francis Group, LLC

Library of Congress Cataloging-in-Publication Data (applied for)

ISBN: 978-1-032-59806-2 (hbk)
ISBN: 978-1-032-59808-6 (pbk)
ISBN: 978-1-003-45634-6 (ebk)

DOI: 10.1201/9781003456346

Typeset in Times New Roman
by Prime Publishing Services

Editors

Shoufeng Cao
QUT Design Lab, Queensland University of Technology, Brisbane, Australia
shoufeng.cao@qut.edu.au, https://orcid.org/0000-0001-5178-7454

Dr Shoufeng Cao is an Industry Fellow at the QUT Design Lab, Queensland University of Technology (QUT), and an ARC Research Fellow at the School of Agriculture and Food Sustainability, Faculty of Science, University of Queensland (UQ). He is currently working on the research project *"Deadly Solution: Towards an Indigenous-led Bush Food Industry"*, exploring and demonstrating the transformative potential of blockchain and non-fungible tokens (NFTs) in the booming bushfood industry. His research expertise lies in blockchain for businesses, supply chains, industries, communities and society.

Shoufeng obtained his doctoral degree from UQ, with his thesis on system-wide, data-driven risk analysis and management of global fresh-produce value chains, for the strategic competition of supply chain to supply chain. Shoufeng started his post-doctoral research journey at UQ and QUT, where he completed three digital transformation projects in agri-food supply chains and the industry funded by three industry-led Commonwealth-funded Cooperative Research Centres including the CRC for Developing Northern Australia (CRCNA), the Food Agility CRC, and the Future Food Systems CRC.

His research spans identification of digital transformation strategies in complex multi-industry, multi-region contexts, to the design, implementation, and evaluation of blockchain solutions for end users, industrial transformation and real-world disruptive impacts. Shoufeng's projects have been awarded two Good Design Australia Awards (2020) and have placed as runner-up in the 2021 ACS Digital Disruptors Awards, in the ICT Research Project of the Year category.

Marcus Foth
QUT Design Lab, Queensland University of Technology, Brisbane, Australia
m.foth@qut.edu.au, https://orcid.org/0000-0001-9892-0208

Marcus Foth is a Professor of Urban Informatics at the School of Design, Faculty of Creative Industries, Education, and Social Justice, Queensland University of Technology, Brisbane, Australia. He is also a Chief Investigator in the QUT Digital Media Research Centre (DMRC), and a founding member of the More-than-Human Futures research group. Marcus' research brings together people, places, and technologies. His current research comprises urban media and geoprivacy; data care in smart cities; creating better digital futures; blockchain and food system innovation, and; sustainability and more-than-human futures.

For over two decades, Marcus has led ubiquitous computing and interaction design research into interactive digital media, screen, mobile and smart city applications. Marcus founded the Urban Informatics Research Lab in 2006. He is a Fellow of the Australian Computer Society and the Queensland Academy of Arts and Sciences, a Distinguished Member of the international Association for Computing Machinery (ACM), and currently serves on Australia's national College of Experts (2021 – 2025).

Marcus has authored over 280 peer-reviewed publications. His recent books include *"Designing More-than-Human Smart Cities: Beyond Sustainability, Towards Cohabitation"* (Oxford University Press, 2024) and *"Community, Culture, Commerce: The intermediary in design and creative industries"* (Palgrave Macmillan, 2024).

Marcus' research has received many awards. For example, his BeefLedger project was recognised as a 2020 Good Design Australia Award winner twice, in the commercial services category and for design strategy, for demonstrating tangible benefits and impact of blockchain and smart contract applications for industry. Nominated by the Smart Cities Council Australia / New Zealand, Marcus was awarded ICT Researcher of the Year 2017 (Gold Disruptor) by the Australian Computer Society.

Preface

The origins of blockchain can be traced back to 2008 with the publication of Satoshi Nakamoto's (2008) white paper, "Bitcoin: A Peer-to-Peer Electronic Cash System." This foundational document introduced the concept of a decentralised digital currency, Bitcoin, and laid the groundwork for blockchain technology. Nakamoto described the system as "a system for electronic transactions without relying on trust." The launch of Bitcoin in 2009 marked the beginning of a new era in digital finance, where transactions could occur without the need for a centralised authority, relying instead on cryptographic proof and a distributed network of nodes.

As cryptocurrencies were brand new and thus largely unregulated in those early days, Bitcoin quickly became the main form of payment for pornographic content and drug deals on the "Silk Road" or dark web (Barratt, 2012; Christin, 2013; Foley et al., 2019). The following years saw blockchain's application grow beyond cryptocurrencies, leading to a period of intense hype around 2018 that some have even termed a "revolution" (Tapscott & Tapscott, 2016). During this time, blockchain was heralded as a transformational technology with the potential to disrupt numerous industries. However, this enthusiasm often lacked a critical understanding of the technology's broader implications, particularly regarding its social and ethical dimensions (De Filippi et al., 2024; Powell et al., 2023). The focus was largely on the technical capabilities and financial prospects of blockchain, overshadowing its potential to foster positive societal change.

As initial excitement has settled, there is a growing recognition of the need to consolidate scholarly and industry thinking around the applications of blockchain *for good*. With this book, *Blockchain for Good: The Transformative Impacts on Industry, Community, and the Planet*, we seek to contribute to this conversation. The book explores how blockchain technology can be utilised for ethical, social, and environmental benefits. By examining case studies and theoretical perspectives, it aims to move beyond the technical aspects of blockchain and consider its potential to drive positive change in various domains.

The book is structured into three parts:

- **Industry** – Blockchain for Better Business Practices;
- **Community** – Blockchain for a Fairer and Just Society, and;
- **Planet** – Blockchain for a Sustainable Future.

Each section looks at how blockchain can enhance business practices, support fairer societal systems, and promote sustainable environmental actions, respectively. This interdisciplinary approach, bringing together voices from computer science, information systems, business management, and the humanities and social sciences, aims to provide a comprehensive understanding of blockchain's potential to serve the common good. The goal is to demonstrate that blockchain technology can be more than a financial instrument; it can be a catalyst for ethical and sustainable development in an increasingly digital world.

Our own academic and scholarly journey of exploring blockchain's transformative impacts in practice began in 2017. At that time, we established university-industry partnerships with Brisbane-based blockchain pioneers to co-design industry-based blockchain solutions. Through our first "blockchain for beef traceability" proof-of-concept (PoC) project (2018–2020) funded by Food Agility CRC (Collaborative Research Centre), we explored whether and how blockchain technology can redefine trust in supply chains (Powell et al., 2023). As part of this work, we introduced blockchain-based supply chain solutions to strengthen consumer trust (Cao et al., 2021) and streamline trade finance (Natanelov et al., 2022) in the beef industry. In our following "blockchain and smart trade hub" project funded by Future Food Systems CRC (2020–2022), we expanded our focus to critically examine the cross-border trade (Cao, Boyen, et al., 2022) and supply chain governance (Cao, Foth, et al., 2022), while developing social-technical solutions (Powell et al., 2021) to complement blockchain's inherent limitations in mitigating the garbage in garbage out issue (Powell et al., 2022). We also advanced blockchain solutions for supply chain finance by developing an asset-backed instrument (Miller et al., 2023). Our exploration continued with the support of QUT's Future Leaders Scheme (2021–2022). Through the "blockchain for food sustainability" project, we explored a better way of achieving sustainability communication enabled by blockchain traceability (Cao, Johnson, et al., 2023) and proposed a design architecture and pathway to inform the technical development (Cao, Xu, et al., 2023).

Producing this book is a natural extension of this long-term program of research demonstrating our commitment to leveraging blockchain for positive transformative impacts. We are certainly not alone on this journey; hence with this book we sought to engage with scholars and practitioners across the globe and collate their insights, learnings and studies that strive for blockchain's real-world benefits. By sharing our knowledge and experiences, we hope to inspire the exploration and adoption of blockchain solutions that can drive meaningful changes.

We are pleased to have been able to assemble a diverse and original array of chapter contributions from author teams from across the world. Section 1 focuses on Industry – **Blockchain for Better Business Practices**. Hilde Heim opens this section with Chapter 1, titled *Thread Count: Implementing Blockchain for Better Fashion Supply Chain Transparency*. She explores how blockchain technology can be utilised to enhance transparency and accountability in the fashion industry. The chapter discusses the challenges of integrating blockchain into supply chains, particularly in terms of cost, scalability, and data disclosure. It provides case studies of tech startups working with fashion brands to implement blockchain solutions, looking at the need for both technological advancements and socio-organisational changes to achieve meaningful transformation.

Murat Kizildag, Mahala Geronasso, and Marcos Medeiros have co-authored Chapter 2, *Next-gen Business Architecture with Blockchain in the Service Industry*. They explore the transformative potential of blockchain technology in service-oriented sectors. This chapter examines how blockchain can enhance transparency, security, and efficiency across various service industries, from hospitality to retail, by implementing decentralised systems, smart contracts, and improved data management. Highlighted are the challenges and opportunities of integrating blockchain into business practices, emphasising the potential for innovation and the importance of overcoming barriers to adoption.

Chapter 3, *Empowering Small-Scale Agriculture through Blockchain and Distributed Ledger Technology: Review and Future Perspectives*, written by Sabrina Scuri, Nuno Nunes, and Valentina Nisi, investigates the potential of DLTs to revolutionise the agricultural sector, particularly for smallholder farmers. The chapter discusses the benefits of DLT in enhancing transparency, efficiency, and financial resilience, while addressing the challenges these technologies face in real-world applications. It emphasises the need for a user-centred approach to design DLT solutions that are accessible and beneficial to small farmers, thus promoting more equitable and sustainable agricultural practices.

Małgorzata Macuda, Stessy Coutelle, Rami Alkhudary, and Pierre Féniès have written Chapter 4, *Enhancing Healthcare System Performance Assessment through Blockchain Integration*. They explore the potential of blockchain technology to improve the performance assessment of healthcare systems. The chapter examines how blockchain can address challenges in data security, transparency, and resource management within the Health System Performance Assessment framework. By enhancing the reliability and efficiency of healthcare services, blockchain can foster trust among stakeholders, support patient-centric care, and align healthcare operations with broader societal goals.

Section 2 is devoted to Community – **Blockchain for a Fairer and Just Society**. The first chapter in this section is written by Ryan Bowler, Chris Speed, and Geoffrey Goodell, titled, *Money as a Design Material Towards the Empowerment of Citizens*. In this chapter, the authors consider the evolving nature of money as societies transition from physical cash to digital currencies, specifically focusing on Central Bank Digital Currency (CBDC). The chapter examines the socio-cultural impacts of diminishing physical cash, the design considerations for CBDC, and the potential for digital currencies to reflect diverse user values. It stresses the need for inclusive design to ensure that digital forms of money do not exacerbate existing barriers but instead empower citizens.

Chapter 6, *Blockchain and Tokenisation for Local Communities*, is co-authored by Guido Boella, Claudio Schifanella, and Cristina Viano. This chapter discusses the application of blockchain and tokenisation to foster civic participation and social economies within urban environments. It introduces the concept of Augmented Commoning Areas (ACAs), which blend physical and digital spaces to enable collaborative interactions among citizens, public administrations, and local businesses. The chapter presents the CO3 project's use of blockchain-based wallets, augmented reality, and geolocation technologies to support various forms of economic and social interactions, such as tokenised economies, interactive democracy, and gamification, aiming to enhance local community engagement and co-production of public services.

Chapter 7 follows, entitled, *A Reference Architecture for Blockchain-based Social Good Applications*. Gowri Ramachandran and Raja Jurdak present a comprehensive framework for developing blockchain-based applications aimed at creating social good. The chapter outlines key blockchain properties such as decentralisation, transparency, and immutability, and how these can be utilised in applications ranging from supply chain management to misinformation detection. It discusses the technical challenges and considerations for integrating blockchain with legacy systems and AI algorithms, providing a reference architecture that includes data communication, verifiable computing, and incentive layers.

Chris Speed is back, this time co-authoring Chapter 8 with Chris Elsden and Peter Tilley, titled, *Non-fungibility as a Conceptual Resource for Design*. It explores the concept of non-fungibility as a valuable tool for designers to create multiple forms of value. The chapter critiques the traditional notion of fungibility in economic systems and proposes non-fungibility as a means to incorporate social, ecological, and economic values into design practices. Through a series of design research projects and case studies, the authors illustrate how non-fungible tokens (NFTs) and other digital technologies can be used to challenge existing value constellations and promote more sustainable and regenerative societies.

Section 3 considers our planet as a whole – **Blockchain for a Sustainable Future**. Stessy Coutelle, Rami Alkhudary, and Pierre Féniès begin this section with Chapter 9: *Evaluation of Strategic Resources in a Typical Short Food Supply Chain for Potential Blockchain Implementation*. They assess the strategic resources available in the Plaine de Versailles region's food supply chain to determine readiness for blockchain integration. Using the Resource-Based View (RBV) framework, the chapter identifies key technological, organisational, physical, relational, and expertise assets necessary for enhancing supply chain transparency, traceability, and sustainability. The study concludes that while blockchain offers significant potential benefits, such as reducing intermediaries and improving trust among stakeholders, its successful implementation requires careful alignment of these strategic resources and overcoming current resource constraints.

Next Behzad Esmaeilian, Hao-Yu Liao, and Sara Behdad co-author Chapter 10, *Circular Economy through Blockchain and Data Analytics*. They situate themselves at the intersection of blockchain technology and enterprise analytics to consider how best to promote circular economy practices. The chapter discusses how new sources of data, advanced data collection technologies, and innovative analytical approaches can enhance product lifecycle management and sustainability efforts. Through case studies, it is demonstrated how blockchain can improve transparency, traceability, and accountability in tracking products throughout their lifecycle, and how data analytics can optimise processes for sustainability, address the last mile problem in data verification, and support the implementation of the EU's Digital Product Passport policy.

Chapter 11 by Katrien and Ine Steenmans, and Phillip Taylor is titled *Blockchain's Transformative Potential for Circular Economy Laws and Policies*. The authors explore how blockchain technology can support the transition to a circular economy by addressing legal and policy challenges. The chapter identifies four pathways through which blockchain can facilitate this transition: equitable participation, reframing rights, distributing responsibility, and transgressing boundaries between public and private law. It discusses the technology's potential to enhance transparency, traceability, and accountability in resource and waste management, as well as the challenges and limitations in its implementation, such as regulatory barriers and environmental concerns associated with blockchain's energy use.

Last but not least, Chapter 12, *Blockchain Applications for Carbon Trading—An Overview of Current Developments in China*, by Yijia Hao, Shoufeng Cao, and Yong Xia analyses the application of blockchain technology in carbon trading markets, particularly focusing on China's initiatives. The chapter highlights blockchain's potential to enhance transparency, traceability, and security in carbon transactions, addressing challenges such as insufficient information disclosure and double counting. It provides an overview of China's national and international standards, research and development projects, and pilot experiments in blockchain-enabled carbon trading, demonstrating China's commitment to leveraging technology for a sustainable and transparent carbon market.

We are grateful to Adjunct Professor Warwick Powell who contributed the book's Epilogue where he reflects on the evolving discourse around blockchain technology and its alignment with the concept of "good." He explores the normative assumptions embedded in the idea of blockchain for good, questioning whether blockchain can truly contribute to societal benefits or if the technology's inherent properties, such as anonymity and permissionless participation, complicate its potential for positive impact. Powell critiques the initial optimism and latter realisations about blockchain's limitations in addressing real-world complexities, highlighting the ongoing challenges in achieving meaningful and ethical applications of the technology.

We conclude by acknowledging that blockchain technology continues to evolve, which is why its potential to contribute to societal good remains a compelling area of exploration. Future developments will likely focus on enhancing scalability, interoperability, and accessibility of blockchain systems to address existing challenges and broaden impact, as well as looking at the nexus of technical, organisational and social aspects. Emphasis will also be placed on ethical considerations, ensuring that blockchain applications prioritise transparency, inclusivity, and sustainability. As more industries and communities adopt blockchain, the dialogue around its role in promoting social, environmental, and economic benefits will deepen, guiding the development of frameworks and standards that support a more equitable and just digital economy. This book aims to inspire continued research and innovation, fostering a future where blockchain technology is harnessed responsibly for the greater good.

Shoufeng Cao & Marcus Foth
Brisbane, Australia, August 2024

References

Barratt, M.J. (2012). Silk Road: eBay for drugs. *Addiction (Abingdon, England)*, *107*(3), 683. https://doi.org/10.1111/j.1360-0443.2011.03709.x

Cao, S., Boyen, X., Deane, F., Miller, T. and Foth, M. (2022). Enabling cross-border trade in the face of regulatory barriers to data flow – the case of the blockchain-based service network. In *Blockchain for Industry 4.0* (pp. 285–303). CRC Press. https://doi.org/10.1201/9781003282914-15

Cao, S., Foth, M., Powell, W., Miller, T. and Li, M. (2022). A blockchain-based multisignature approach for supply chain governance: A use case from the Australian beef industry. *Blockchain: Research and Applications*, *3*(4), 100091. https://doi.org/10.1016/j.bcra.2022.100091

Cao, S., Johnson, H. and Tulloch, A. (2023). Exploring blockchain-based Traceability for Food Supply Chain Sustainability: Towards a Better Way of Sustainability Communication with Consumers. *Procedia Computer Science*, *217*, 1437–1445. https://doi.org/10.1016/j.procs.2022.12.342

Cao, S., Powell, W., Foth, M., Natanelov, V., Miller, T. and Dulleck, U. (2021). Strengthening consumer trust in beef supply chain traceability with a blockchain-based human-machine reconcile mechanism. *Computers and Electronics in Agriculture*, *180*, 105886. https://doi.org/10.1016/j.compag.2020.105886

Cao, S., Xu, H. and Bryceson, K. P. (2023). Blockchain Traceability for Sustainability Communication in Food Supply Chains: An Architectural Framework, Design Pathway and Considerations. *Sustainability*, *15*(18), 13486. https://doi.org/10.3390/su151813486

Christin, N. (2013). Traveling the silk road: a measurement analysis of a large anonymous online marketplace. *Proceedings of the 22nd International Conference on World Wide Web*, 213–224. https://doi.org/10.1145/2488388.2488408

De Filippi, P., Reijers, W. and Mannan, M. (2024). *Blockchain Governance*. MIT Press.

Foley, S., Karlsen, J.R. and Putniņš, T.J. (2019). Sex, Drugs, and Bitcoin: How Much Illegal Activity Is Financed through Cryptocurrencies? *The Review of Financial Studies*, *32*(5), 1798–1853. https://doi.org/10.1093/rfs/hhz015

Miller, T., Cao, S., Foth, M., Boyen, X. and Powell, W. (2023). An asset-backed decentralised finance instrument for food supply chains – A case study from the livestock export industry. *Computers in Industry*, *147*, 103863. https://doi.org/10.1016/j.compind.2023.103863

Nakamoto, S. (2008). *Bitcoin: A peer-to-peer electronic cash system*. https://doi.org/10.2139/ssrn.3440802

Natanelov, V., Cao, S., Foth, M., & Dulleck, U. (2022). Blockchain smart contracts for supply chain finance: Mapping the innovation potential in Australia-China beef supply chains. *Journal of Industrial Information Integration*, 30, 100389. https://doi.org/10.1016/j.jii.2022.100389

Powell, W., Cao, S., Foth, M., He, S., Turner-Morris, C. and Li, M. (2023). Revisiting Trust in Supply Chains: How Does Blockchain Redefine Trust? In A. Bouras, I. Khalil, & B. Aouni (Eds.), *Blockchain Driven Supply Chains and Enterprise Information Systems* (pp. 21–42). Springer International Publishing. https://doi.org/10.1007/978-3-030-96154-1_2

Powell, W., Cao, S., Miller, T., Foth, M., Boyen, X., Earsman, B., del Valle, S. and Turner-Morris, C. (2021). From premise to practice of social consensus: How to agree on common knowledge in blockchain-enabled supply chains. *Computer Networks*, *200*, 108536. https://doi.org/10.1016/j.comnet.2021.108536

Powell, W., Foth, M., Cao, S. and Natanelov, V. (2022). Garbage in garbage out: The precarious link between IoT and blockchain in food supply chains. *Journal of Industrial Information Integration*, *25*, 100261. https://doi.org/10.1016/j.jii.2021.100261

Tapscott, D. and Tapscott, A. (2016). *Blockchain revolution: how the technology behind bitcoin is changing money, business, and the world*. Penguin. https://doi.org/10.5555/3051781

Acknowledgements

We would like to extend our heartfelt gratitude to all the chapter authors who have contributed their valuable insights and expertise to this book as well as contributed their time and expertise as peer reviewers, as well. Your hard work and dedication have been instrumental in bringing this project to fruition.

Each chapter has been peer reviewed by the editors, by another chapter author, and by an external reviewer. We thus also sincerely thank our external peer reviewers for their meticulous evaluations and constructive feedback, which have significantly enhanced the quality of the content and strengthened the rigour and coverage of each chapter:

- Arlene Bailey, University of the West Indies, Jamaica
- Kamanashis Biswas, Australian Catholic University, Australia
- Margaret Bruce, Manchester Metropolitan University, UK
- Felicity Deane, Queensland University of Technology, Australia
- Yu Gong, University of Southampton, UK
- Greg Hearn, Queensland University of Technology, Australia
- Sara Heitlinger, City University of London, UK
- Marko Hölbl, University of Maribor, Slovenia
- Babu Pillai, Southern Cross University, Australia
- Lachlan Robb, Queensland University of Technology, Australia
- Maurizio Teli, Aalborg University, Denmark
- Anoop V. S., Kerala University of Digital Sciences, Innovation and Technology, India

Our appreciation goes out to the commissioning editor Vijay Primlani at CRC Press and all publishing support staff for their assistance and guidance throughout the publishing process. Your expertise and commitment have been invaluable in shaping the final outcome of this book.

We are deeply grateful to our colleagues for their encouragement and support during the development of this book. Your insights, discussions, and camaraderie have been a source of inspiration and motivation.

Lastly, we thank our families for their patience and understanding, allowing us to dedicate the time and effort needed to complete this project.

List of Abbreviations

ACA	:	Augmented Commoning Areas
AI	:	Artificial Intelligence
BCER	:	Beijing Forest Carbon Emissions Reduction
CBDC	:	Central Bank Digital Currency
CCER	:	China Certified Emission Reduction
CDM	:	Clean Development Mechanism
CQEA	:	Chongqing Emission Allowance
DLT	:	Distributed Ledger Technology
DAO	:	Decentralised Autonomous Organization
ETH	:	Ethereum
EU	:	European Union
EU ETS	:	European Union Emissions Trading System
FJEA	:	Fujian Emission Allowance
FFCER	:	Fujian Forestry Certified Emission Reduction
GDEA	:	Guangdong Emission Allowance
GHG	:	Greenhouse Gas
HBEA	:	Hubei Emission Allowance
HSPA	:	Health System Performance Assessment
IoT	:	Internet of Things
MIIT	:	Ministry of Industry and Information Technology
NFT	:	Non-Fungible Token
PHCER	:	Pu Hui Certified Emission Reductions
RBV	:	Resource-Based View
R&D	:	Research and Development
SHEA	:	Shanghai Emission Allowance
SZEA	:	Shenzhen Emissions Reduction
TJEA	:	Tianjin Emission Allowance
YPG	:	Yunnan Power Grid Corporation

Contents

List of Contributors

Rami Alkhudary

Rami Alkhudary, PhD, is an Assistant Professor at Université Paris-Panthéon-Assas, France, where he earned his PhD in Management Science, focusing on blockchain technology in supply chains. He received the university's Thesis Award in Management Science in 2021. His research centres on emerging technologies in supply chain and project management. Rami's work has been recognised at international conferences, winning the Best Research Paper Award at PROLOG 2019. His publications appear in journals such as the International Journal of Production Research, International Journal of Project Management, Marketing Letters, and Harvard Business Review France, among others. He is also a reviewer for international journals and conferences.

Sara Behdad

Sara Behdad is an Associate Professor at the Engineering School of Sustainable Infrastructure & Environment at the University of Florida (UF), USA. She holds a PhD in Industrial and Enterprise Systems Engineering from the University of Illinois at Urbana-Champaign. Currently, she directs the Green Engineering Technology for the Community of Tomorrow (GETCOT) lab at UF. Her expertise include product lifecycle engineering, sustainable design, remanufacturing, and decision analysis.

Guido Boella

Guido Boella is a professor at the Department of Computer Science and Vice-Rector for relations with industries and industrial innovation at the University of Turin, Italy. Guido is the author of over 300 international publications. His research considers AI and Blockchain. He is the co-founder of the Italian Society on Ethics and AI (SIpEIA).

Ryan Bowler

Dr Ryan Bowler completed his doctoral research on utilising Temporal Uncertainty Tools to enhance inclusivity in Human-Computer Interaction. Through his dedication to social inclusion and technology, he has developed a fascination with the area of financial inclusion. More specifically, he is intrigued by the way emerging financial technologies shape social dynamics and interconnected structures linked to diverse forms of money. His primary focus is the establishment of barriers around monetary access and choice;

such barriers resulting in social exclusions by misrepresenting diverse values that individuals associate with money and its use beyond numerical depiction. His passion for finance, technology, and inclusion has led him to become a user researcher in the Scottish Government. His role involves assisting in the development of a user-centric payments platform, while also exploring inclusion and payments.

Stessy Coutelle

Stessy Coutelle is a second-year PhD candidate in Management Science at Université Paris-Panthéon-Assas, France. She holds dual master's degrees from Université Paris-Panthéon-Assas in Supply Chain Management, from PPA Business School, and Consulting and Research in Organisations, specialising in logistics project management. Her research is focused on integrating blockchain technology into sustainable supply chains. Passionate about technological innovations and their impact on traceability and operational efficiency, Stessy aims to enhance transparency and promote sustainability within value chains, providing valuable insights into the field.

Chris Elsden

Dr Chris Elsden is a Chancellor's Fellow in Service Design at the Institute for Design Informatics, University of Edinburgh, UK. He is a design-researcher, with a background in sociology and expertise in the human experience of data-driven services. Using and developing innovative design-research methods, his work undertakes diverse, qualitative and often speculative engagements with participants to investigate emerging relationships with technology—particularly data-driven tools, FinTech and blockchain technologies.

Behzad Esmaeilian

Behzad Esmaeilian is an Assistant Professor at the Andrew F. Brimmer College of Business and Information Sciences, Tuskegee University, Alabama, USA. He holds an MS in Industrial and Systems Engineering from the University of Florida and a PhD in Industrial Engineering and Management Systems from the University of Central Florida. His research interests include sustainable supply chains, blockchain technology, and sustainable business strategies.

Pierre Féniès

Pierre Fenies, PhD, HDR, is a Professor of Logistics and Supply Chain Management at Université Paris-Panthéon-Assas. His expertise includes control management, logistics, supply chains, and business intelligence. Pierre's research and teaching activities have enhanced software solutions and decision-making processes through computer technologies. He partnered with Braincube SAS (formerly IP Leanware) to develop a business intelligence solution for industrial and general management control. His research has been validated in various supply chains, including industrial and medical sectors, and has been published in numerous journals.

Mahala Geronasso

Mahala Geronasso is a PhD candidate at the Rosen College of Hospitality Management, University of Central Florida (UCF), USA. Specialising in hospitality finance and technology, her research utilises advanced statistical techniques to enhance financial performance in the hospitality industry. Her teaching experience includes managerial accounting, revenue management, and hospitality management/leadership fundamentals. Mahala's research interests encompass AI, blockchain, finance, and management. She holds an MBA and a Bachelor's degree in Hospitality and Tourism Management from UCF and has extensive leadership experience in the hospitality and tourism industry.

Geoffrey Goodell

Dr Geoffrey Goodell is a Lecturer of Financial Computing at University College, London, whose research focuses on socio-technical systems in financial services. He is a member of the Bank of England CBDC Technology Forum and an associate of the LSE Systemic Risk Centre. He routinely contributes to international standards and advises industry groups on digital assets technology and policy. Dr Goodell is a CFA charterholder and previously served as a strategist at Goldman Sachs & Co and as a partner of a boutique asset management firm. His research advances knowledge and policy at the intersection of finance, technology, and regulation.

Yijia Hao

Hao Yijia is a PhD candidate at Xinjiang Agricultural University, China, majoring in agricultural and forestry economic management. Her previous research interests are related to food security and production factor allocation. She has published a monograph and several academic papers in this field and participated in one National Natural Science Foundation of China and three provincial fund projects. Her current research area is grassland carbon sinks and carbon market trading.

Hilde Heim

Dr Hilde Heim is a Senior Lecturer and the Deputy Division Head of Fashion Communication at the Manchester Fashion Institute, Manchester Metropolitan University, UK. Through her experience as a fashion entrepreneur, Hilde explores the nexus of supply chain transparency, emerging technologies, and global compliance, and their effect on small to medium enterprises (SMEs). Her work investigates blockchain, AI, smart tagging, and digital product passports. In connection with business, NGOs, trade associations and multi-stakeholder initiatives (MSIs), she aims to build knowledge on sustainable and ethical supply chain collaboration in the fashion industry.

Raja Jurdak

Raja Jurdak is a Professor of Distributed Systems, Chair in Applied Data Sciences, and Director of the Trusted Networks Lab at Queensland University of Technology. His research interests include trust, mobility, security, and energy-efficiency within networks. He has authored over 260 publications, including 3 books, which have collectively garnered over 15,000 citations. Prof. Jurdak serves on the editorial boards of IEEE Transaction on Network and Service Management and Ad Hoc Networks. He is an Adjunct Professor with the University of New South Wales, a Senior Member of the IEEE and a Distinguished Visitor of the IEEE Computer Society.

Murat Kizildag

Dr Murat Kizildag is an Associate Professor of Finance at the Rosen College of Hospitality Management, University of Central Florida, USA. Dr Kizildag is intrigued by the depth of financial research. His research expertise spans the areas of Security Valuation, Risk & Investment Analytics, Financial Modeling & Forecasting, Portfolio Management & Optimisation, and Time-Series Analysis. Dr Kizildag has produced more than 50 academic articles and attended numerous academic and industry conferences. His recent articles have appeared in business finance and hospitality business/ management journals. Dr Kizildag has worked on several different grant projects and contracts.

Hao-yu Liao

Hao-yu Liao is a Postdoctoral Associate at the University of Florida's Institute of Food and Agricultural Sciences, USA. He holds a PhD in Environmental Engineering Science from the University of Florida and a Master's degree in Bioenvironmental Systems Engineering from National Taiwan University. His research interests include machine learning, optimisation methods, and data analysis, with applications in environmental studies such as flood modelling and ecosystem service assessments.

Małgorzata Macuda

Małgorzata Macuda, PhD, is an Assistant Professor in the Department of Management Accounting at the Poznań University of Economics and Business, Poland. She completed her postgraduate studies in health care management at the Poznań University of Medical Sciences. She is a Visiting Professor at Universidad Bernardo O'Higgins Centro de Investigación Institucional, Chile, and at ESFAM, Bulgaria. Her research interests include financial accounting, cost accounting, management accounting, hospital accounting, and sustainability accounting. She is an author and a co-author of more than 60 scientific publications and the Deputy Editor-in-Chief of the international journal Revue Internationale des Économistes de Langue Française (RIELF).

Marcos Medeiros

Dr Marcos Medeiros is an Assistant Professor of Hospitality & Finance at the University of Virginia's College at Wise, USA, specialising in finance, real estate, and hospitality. His research focuses on the nexus of real estate and hospitality, exploring the dynamics of vacation rental performance, valuation, and leverage. Dr Medeiros has a keen interest in the transformative potential of emerging technologies, including blockchain, artificial intelligence, and mobile applications, and how these advancements can revolutionise real estate transactions and financial transparency. With over 30 years of experience in the investment banking sector, Dr Medeiros has led major capital-raising initiatives across various international markets, including Brazil, Spain, France, and the USA.

Valentina Nisi

Valentina Nisi is Professor of Design Thinking at the Instituto Superior Técnico (IST), University of Lisbon, Portugal, where she teaches and researches digital interactive design, art, and media. Valentina has published, exhibited, and taught internationally in digital media, art, and design. Her research focuses on designing and producing digitally mediated experiences in controlled spaces and "in the wild," merging architecture, environment, landscape, multimedia, narrative, and artistic elements.

Nuno Nunes

Nuno Nunes is Professor of Human-Computer Interaction at the Instituto Superior Técnico (IST), University of Lisbon, Portugal, and the President of the Interaction Technologies Institute (ITI). He is also co-director of the Carnegie Mellon International Partnership and visiting faculty at the Human-Computer Interaction Institute at Carnegie Mellon University, USA. Nuno was the main proponent of the Bauhaus of the Seas – New European Bauhaus lighthouse project. His research interests lie in applying models to software, systems, and services for environmental sustainability and participatory culture. Nuno is a strong advocate of the role of design in engineering.

Warwick Powell

Warwick Powell is an Adjunct Professor in the QUT Design Lab at Queensland University of Technology. He is a global thought leader in blockchain design strategy with peer-reviewed publications in leading journals. He is the author of the monograph, *"China, Trust and Digital Supply Chains: Dynamics of a Zero Trust World"* (Routledge, 2023). Warwick is the founding chairman of *Smart Trade Networks*, a blockchain technology development and commercialisation company as well as *BeefLedger*, a global leader in applied blockchain solutions to the red meat sector. He has served as policy advisor to The Hon Dr Kevin Rudd AC, former Australian Prime Minister. He continues to teach professional courses in areas such as innovation, creativity, regional economic development and blockchain technology and has delivered guest lectures to undergraduate and masters level students at Queensland University of Technology, Jiaotong University (Shanghai), Beijing Normal University, Shenzhen University, James Cook University and Edith Cowan University.

Gowri Sankar Ramachandran

Dr Gowri Sankar Ramachandran is a Senior Lecturer at the School of Information Systems, Queensland University of Technology (QUT), specialising in cybersecurity. He earned his PhD in Computer Science from KU Leuven, Belgium, and has held research positions at QUT, Australia, the University of Southern California, USA, and KU Leuven, Belgium. His work has earned him many awards, including four best paper recognitions from conferences and an Honorary Certificate of Appreciation from the IEEE Communication Society. Gowri is on the review board of IEEE Transactions on Services Computing and has organised several international conferences, including ICBC, ICSOC, SenSys, and IPSN.

Claudio Schifanella

Dr Claudio Schifanella is an Associate Professor at the Department of Computer Science, University of Turin, Italy. His research topics include civic technologies, social computing, distributed ledger technologies, data mining, multidimensional data analysis, knowledge representation, and text mining.

Sabrina Scuri

Sabrina Scuri is an Assistant Professor at the Department of Design, Politecnico di Milano, Italy, and a member of the Interactive Technologies Institute/Laboratório Associado de Robótica e Sistemas de Engenharia (ITI/LARSyS). Her expertise reside at the intersection of interaction design, user experience, and human-computer interaction. In particular, her research focuses on designing interactive systems to help promote behaviour change and inform sustainable development initiatives in all domains of sustainability – environmental, social, and economic.

Chris Speed

Chris Speed, FRSE, FRSA, is Professor of Design for Regenerative Futures at RMIT, Melbourne, Australia, where he collaborates with a wide variety of communities and partners to explore how design provides methods to adapt toward becoming a regenerative society. Chris has an established track record in directing large complex grants and educational programmes with academic, industry and third sector partners, that apply design and data methods to social, environmental and economic challenges.

Ine Steenmans

Dr Ine Steenmans is Associate Professor of Futures, Analysis, and Policy at the University College London (UCL), UK. Ine's research focuses on future policy competencies and capabilities and asks how their development can be better aligned. She has a special interest in the competencies for integrating knowledge across sectors and time horizons. This includes futures and foresight methods, systems methodologies, and evaluation. All her work takes the same needs-led, transformation-oriented, and partnership-based approach. Before joining UCL, Ine worked as a foresight- researcher at the UK Government Office for Science.

Katrien Steenmans

Dr Katrien Steenmans is a postdoctoral researcher on the Circular Supply Chains (CirCus) project at the Faculty of Law, University of Copenhagen, Denmark. This project investigates the legal risks of adopting extended producer responsibility and liability within private law and waste law to promote circular economies. Katrien's previous research has focused on environmental law and in particular waste law, including the role of waste regulations and policies in enabling industrial symbiosis; implications of different property rights in waste; and how blockchain technology can support waste regulations and policies.

Phillip Taylor

Dr Phillip Taylor is an honorary research fellow at the Department of Computer Science, University of Warwick, UK. Following his PhD in data mining of vehicle telemetry in 2015, Phillip undertook postdoctoral research comprising several areas, including agent-based systems, trust and reputation, data compression, and vehicle location privacy. Phillip then joined his industry and worked as a data scientist at Trilateral Research Ltd, where he worked with large language models, knowledge graphs, and geospatial data. He currently works as a software engineer in the finance sector.

Peter Tilley

Peter Tilley is a freelance designer and illustrator in Edinburgh, UK. He has a keen interest in digital, often interactive, media and material. A background in traditional painting and drawing supplements Peter's work in digital mediums. His clients often seek to blend the values of traditional illustration with technical ideas and modalities for a variety of audiences.

Cristina Viano

Dr Cristina Viano is a postdoctoral researcher at the Department of Computer Science, University of Turin, Italy. Her research interests consider digital geography, with a focus on the socio-spatial implications of civic technologies in local communities, and experimentation with blockchain-based platforms.

Yong Xia

Xia Yong is a professor and doctoral supervisor at Xinjiang Agricultural University, China and a member of the board of directors of Xinjiang Agricultural Economics Society. His main research interests include rural finance, regional economy, and international trade of agricultural products. He has presided over more than 20 scientific research projects, including a National Natural Science Foundation of China and a National Social Science Fund of China. He has published over 100 papers, edited 1 monograph, and participated in 3 compilation books.

1 Thread Count: Implementing Blockchain for Better Fashion Supply Chain Transparency

Hilde Heim[1*]

"Technology is essentially the prime enabler that allows sustainable fashion to thrive and develop today" (Scaturro, 2008).

1. Introduction

The provision of more accurate and verifiable supply chain transparency (SCT) is pressing given the untenable practices in the fashion industry that are 'contributing directly and significantly to the triple planetary crisis of climate change, nature and biodiversity loss, pollution and waste' (Arthur, 2023). The UN reports that in 2022—7 years away from the targets set—the industry was not on track to meet climate action goals (Geldard and Ellerbeck, 2023). This points to the urgency, and indeed acceleration required to meet objectives – an aspiration that can only be achieved through technological intervention (Cole et al., 2019). Technology provides efficiencies beyond the 'DIY' data entry capabilities of spreadsheets, product lifecycle management (PLM) systems and/or self-declaration on in-house websites. Thousands, if not millions, of data points require handling to achieve comprehensive SCT (Bain, 2024). According to Amed et al. (2019), 75% of retailers plan to invest in technology in the future, and 85% of corporations are already experimenting with technology (Chui et al., 2023; PwC, 2018). Yet, advanced information communication technology such as blockchain has seen lukewarm reception in the fashion industry. Gartner (2019) explains this as a 'normal' phenomenon in the 'hype cycle' of technology adoption. Even so, all firms—even micro firms—are already held accountable in some way, required to report and verify their business information for compliance (e.g., for income tax reporting obligations). Sustainability reporting is and will be regulated in the near future. Regardless of state level legislation, compliance will be mandated through cross-border flows and therefore affect businesses world-wide (UN/CEFACT, 2021). Beyond the already drafted, tabled and enacted laws of the mid 2020s, more stringent rules will continue to be introduced if climate targets are to be reached—possibly including added taxes on virgin fibre or fossil fuel use (EllenMcArthurFoundation, n.d.; Bounds and Hancock, 2023). These factors

[1] Manchester Fashion Institute, Manchester Metropolitan University, UK.
* Corresponding author: h.heim@mmu.ac.uk

combined, point to the need for a system that has global coordination capabilities, and that can help address the 'hard problems' with which the world is presently confronted, including the disclosure of reliable, science-based measurement data and exposure that holds supply chain stakeholders accountable for their sourcing, production and end-of-life processes.

1.1 The Fashion System

Fashion retailers have long been accused of 'chasing cheap labour across the globe' while 'failing to pay their workers a living wage, using child labour, turning a blind eye to abuses of human rights, being complicit with repressive regimes in denying workers the right to join unions, and failing to enforce minimum labour standards in the workplace' (Maitlandet et al., 2003, p. 597). Easily accessible, cheap off-shore labour markets led to the emergence of fast fashion in the 1990s, which in turn prompted a significant increase in demand from consumers. This cycled into pressure on retailers to minimise costs. Fashion sales doubled between 2010 and 2024, but the frequency of wearing an item decreased by 40% (EllenMacArthurFoundation, 2024). Typically, most garments are discarded after only 7 to 10 wears: around 73% are incinerated or discarded into landfills, while less than 1% are recycled for re-use. Furthermore, the fast fashion business model has resulted in the development of extensive, intricate, and impenetrable supply chain information. Although harmful, and despite numerous global initiatives, the fast fashion business model has not been impeded. Instead, it has escalated (Niinimäki et al., 2020). Within this exploitative landscape, private standards, such as retailers' ethical statements, environmental-social-governance (ESG), or multi-stakeholder initiatives, like the Ethical Trading Initiative (ETI), arose to manage corporate social responsibility (CSR). This was partly because of the lack of universal regulations, and partly because of firms' tokenistic attempts to be seen as more trustworthy (Christopherson and Lillie, 2005; Tallontire, 2007). However, these self-proclaimed 'codes of conduct' were only part of the change needed. Other levers for systems change that emerged subsequently include circularity, alternative value recognition—e.g., the triple bottom line or 3BL (Elkington, 1998)—and transparency.

Currently, transparency, that is, access to the information on producers, processes and products that flow between and beyond firms, is fragmented and uncoordinated. This means that on the one hand firms cannot be sure of—or have no access to—information from their suppliers (and their suppliers' suppliers). On the other hand, they cannot share and verify information adequately, even if motivated to do so. Some altruistic organisations have already taken on the challenge of implementing advanced technologies (including blockchain) to foster improved fashion business practice. For example, in conjunction with their comprehensive rating approach, the observer organisation *Good On You* provides the fashion ecosystem with enhanced access to detailed, accurate, and reliable fashion business data. Other transparency organisations that operate as *quasi*-directories for fashion suppliers globally include Open Source Map and Open Supply Hub (formerly Open Apparel Registry). The UNECE has a blockchain enabled '*Transparency Toolbox*'[1] which is globally accessible. Access to this information, in turn, enables fashion brands first to ensure ethical sourcing and second to furnish more refined ratings, and craft more captivating consumer facing narratives surrounding their products. However, these platforms only offer data on 'economic operators' – that is the suppliers and factories. The data is still fragmented, has no standard methodology and is mostly incapable of individual item identification. Although factory and brand level information is a good start, it is only through the end-to-end tracking and tracing of individual garments (and their numerous components) that we can further systems change. Once the components are tagged, the data behind their origin, composition, movement, recyclability, and end-of-life measures, amongst other elements will require storage, validation and accessibility.

1.2 Communicating and Reporting, Certifications, and Auditing

Until now, standards, certifications and other sustainability credentials (e.g., GOTS, BCI, OEKO-TEX[2]) have served the purpose of providing third-party validation of sustainability (and other) measures and claims. Standards are sets of measurements designed to uphold defined qualities within industry sectors. Certifications validate that standards have been followed. Certification organisations collect evidence by conducting life cycle assessments (LCAs), onsite audits and/or digital data collection (e.g. energy use). Credentials are either generic (e.g., Fair Trade) or specific to a purpose, function or origin (e.g., Forest Stewardship Council – FSC). This means that firms need to consider, and often select an assortment of, standards/certifications to cover the claims they wish to verify. However, certifications are under examination—at best because of their mostly one-dimensional, self-reporting methodologies and at worst because they can conceal exploitation (Mellick et al., 2021; Keniry, 2020). The deteriorating trust in audits, certifications and agencies has led to demand for improved verification and science-based data. Added to the sheer scale of data sharing required, many fashion stakeholders have turned to investigating the capabilities of advanced technology.

1.3 The Opportunity Provided by Blockchain

Capturing vast amounts of varied and detailed data and thereby promising to remedy some of the fashion industry's ills, Web3 enabled digital solutions and associated applications have proliferated since 2014[3] (Behr, 2018; Bak, 2019; Radocchia, 2018; Sargeant, 2017; Swan, 2015). Unlike current, static, centralised

---

1 https://thesustainabilitypledge.org/toolbox.html

2 GOTS: *Global Organic Textiles Standard;* BCI: *Better Cotton Initiative; OEKO-TEX Standard 100*

3 …when Ethereum co-founder Wood described it as a "decentralized online ecosystem…" Edelman, G. (2021) 'The Father of Web3 Wants You to Trust Less', Wired. Available at: https://www.wired.com/story/web3-gavin-wood-interview/ 2024]...

Web2 platforms, Web3 is a persistent online space with interactive capabilities. One of the fundamental constituents of Web3 is blockchain. Blockchain is a stacked software infrastructure that enables several functions. It is used to store, secure and selectively share information held on a distributed network. The various forms of data, include transactions that are particularly relevant for cryptocurrencies, as well as events such as the deployment of smart contracts (Natanelov et al., 2022). The data can encompass a wide range of objects and entities, including but not limited to products, materials, and suppliers. When combined with smart tag tracking (data carriers), blockchain allows users to trace products through their entire lifecycle, encompassing the journey from the first raw material stage through production, distribution, consumption, and end of life. The co-ordination of these related technologies that offer product identification through data capture, storage and sharing has been dubbed the ‘digital product passport’ (DPP) (CIRPASS, 2023).

Beyond coordination for transparency, Warren et al. (2019) discuss the economic advantages and other forms of new value creation offered by blockchain including incentivisation and cross-sector communication. Dormant economic advantages include savings in time, risk aversion and efficiencies gained through reducing repetitive administration. Berg, Davidson and Potts (2019) reason that the cost of business transactions can effectively be reduced through implementing blockchain, thus offsetting the initial investment and opening a pathway to improved coordination among stakeholders. Some observers herald blockchain technology as the *New Internet* (O’Dair, 2019; Tapscott and Tapscott, 2016; Mainelli, 2017). However, blockchain also has its critics. Some studies point out the difficulty in understanding and adoption and are calling for ease of transfer, interoperability, and open access (Fitzgerald, 2006b; Von Hippel, 2005; Openlink, 2018). Others question the usefulness of the application and doubt the return on investment (Remington, 2020; Ley and Martínez-Pardo, 2020; Lay, 2018). All suggest the technology is still maturing and has not yet reached its full potential (Gartner, 2019; Bullón Pérez et al., 2020). Despite its powerful attributes and numerous applications, blockchain’s software architecture is barely understood outside the tech world. Indeed, the consensus among computer scientists is that blockchain (and Web3) applications remain ‘difficult’ and highly specialised within the IT world (Destefanis et al., 2018; Volpicelli, 2018; van Haaren-van Duijn et al., 2022). The table below outlines the main challenges of the blockchain enabled digital ecosystem. This study selects three of the most pressing challenges: the intra-organisational barrier of cost; the systems barrier of scalability, and the inter-organisational barrier of non-disclosure. These were chosen as they persist as the most frequently debated challenges on the eve of transparency legislation enforcement (CIRPASS, 2023).

1.4 Why Blockchain for Fashion?

The implementation of advanced technologies such as blockchain has the potential to enhance the operational efficiency and effectiveness and (importantly) scale

Table 1 Barriers to technology adoption for fashion supply chain transparency

Stakeholder barriers • Suppliers' lack of tech knowledge • Suppliers' lack of access to infrastructure: software, platforms • **Resistance to sharing data** • Lack of capability to integrate • Resistance to collaboration	**External barriers** • Perceived potential for corruption • Disinterest, too complex • Sceptical about return on investment • No guarantee of alignment with regulatory compliance
Organisational barriers • **Lack of financial and human resources** • Lack of management commitment and support • No organisational strategy for technology adoption • Lack of technical knowledge and expertise - digital hesitancy • Organisational culture resisting change, technology adoption and/or sustainability values	**Systems barriers** • **Few examples at scale** • Concerns regarding privacy and security • Lack of interoperability • Negative public perception of blockchain • Tech immaturity • Digital harms – cost of energy and emissions

of supply chain data collection, storage and sharing. Beyond some studies into blockchain's use in textile supply chains and sustainability (Agrawal et al., 2021; Caldarelli et al., 2021; Francisco and Swanson, 2018), research within the fashion sector is limited, outside verifying origin and authenticity for the purposes of alleviating the counterfeiting of luxury goods (Kemp, 2018; Meraviglia, 2018). Few studies focus on Web3's utility for supply chain transparency, and relatedly, its adoption among under-resourced small-scale fashion firms. Many scholars present the potential for blockchain technology, but few empirical investigations of use cases, such as by Benstead et al. (2024) have been undertaken—likely due to the current lack of such cases. Researchers (Morlet et al., 2017; Patel et al., 2018; Queiroz et al., 2019) concur that as a complex technology with a multitude of applications, blockchain warrants more first-hand research. Building on prior works (Rusinek et al., 2018; Francisco and Swanson, 2018; Agrawal et al., 2021), this study not only investigates the effect of blockchain and related applications on fashion firms, but the shift in the current paradigms that its arrival has instigated, and the effects that may yet occur. One of the key advantages of blockchain is that it can be used as a universal technology of coordination, offering visibility on potentially limitless data points globally (Rennie et al., 2019). This represents a significant systems change to several of the dominant fashion industry conventions of manually entered data, fractured information, trade secrets, and covert activities unethically manipulated for competitive advantage. The introduction of blockchain technology may be a catalyst for positive systems transformation in generating not only economic, but greater environmental and social value. This could be achieved by increased disclosure, awareness, coordination, and an amplified expectation of validated integrity among fashion firms and consumers, thus opening the prospect of alternative business models and management strategies (Casciani et al., 2022; Benstead et al., 2024).

1.5 Small to Medium Fashion Enterprises

The fashion supply chain is replete with small scale firms. Fashion SMEs are found globally with most 'independent designer labels' found in the global north supplied either directly or indirectly by small-scale manufacturers in the global south. In its focus on sustainable supply chain management, the United Nations Economic Commission for Europe (UNECE) is particularly concerned with SMEs as they constitute the majority of organisations under its remit from small scale cotton growers, to outworkers, to specialist craftspeople in developing economies (UNECE, 2023). However, the scope of this study is narrowed to the small-scale fashion brand in the global north. This fashion SME can be a designer label, a small-scale manufacturer, or a supplier of goods and/or services to other fashion firms. It is firms from the global north that are under the most stringent scrutiny from industry observer organisations such as Fashion Revolution, NGOs, consumers and now increasingly from regulators (Arthur, 2023). New knowledge in this sector, including this study, will determine the potential extent of blockchain adoption and how it may generate improved processes and innovations, and ultimately systemic change for the better. This study hypothesises that socio-organisational effects may occur whether tech is implemented or not; that the inter- and intra-firm shifts required in the drive for greater transparency will themselves be of value. This shift in industry dynamics may provide the foundation needed for a smooth transition—whereby the adoption of the tech may occur later, and in a relatively seamless manner once stakeholder relationships have changed for the better. This study suggests that advanced technology should be used first to improve brands' 'hard' profit KPIs such as business operations in economic and operational terms, and second, that what might be perceived as the 'soft' socio-organisational KPIs will follow in due course. This is potentially the better 'sell' by tech to the fashion industry to encourage adoptions and ultimately transformation.

2. Methodology

Many previous studies have investigated the adoption of advanced technologies from the viewpoint of the brand coupled with the way blockchain redefines trust altogether (Powell et al., 2022). Insufficient availability of (economic and human) resources, scalability and a deficit in trust have been identified as significant variables in several blockchain studies (Heim et al., 2022; Openlink, 2018; Fitzgerald, 2006a; Gartner, 2019; Kumar et al., 2017). As a point of departure, and extending on the work of Benstead et al. (2024), this study investigates the challenges faced by tech firms in engaging with SMEs – taking a more detailed view of the three pain points of cost, scalability and non-disclosure. Although it is widely assumed that tech start-ups have complete command of technology—and their proprietary solutions – surprisingly, many solutions providers are themselves still experimenting with technological development. This study investigates the tech start-up's viewpoint of the digital transformation journey by taking a qualitative approach based on interview and case study methodology.

Scholarly literature suggests that utilising case study methodology is advantageous when there is a requirement to gain a comprehensive understanding of a particular issue or phenomenon within its authentic real-world setting (Yin, 2013; Crowe et al., 2011; Stake, 1995). Furthermore, the use of case studies has been extensively embraced in the supply chain literature, demonstrating the ability to offer valuable insights into operational practices within certain sectors, such as the fashion industry (Ye and Lau, 2018; Mao et al., 2007). Through desktop and trade fair enquiry, 80 technology companies were identified that provide Web3 solutions. Ten technology companies were approached (at trade fairs, via associations and/or from media reports) for interview but declined as they were interested in potential commercial relationships rather than research. Two technology startups that have worked with small and mid-sized firms agreed to participate in the research.

In-depth, semi-structured interviews were undertaken with the founders of the 2 startup tech companies, anonymised and identified as DPP solutions provider A (based in Sweden) and DPP solutions provider B (based in the UK). The initial interviews lasted approximately 60 minutes each, and covered several aspects of the startups' vision such as the inspiration behind the development of apps, pilot studies conducted, problems encountered, solutions pursued, and the envisioned future of blockchain for supply chain transparency. Subsequently, the research transitioned to a less formal but more probing approach, deploying methods such as email correspondence, online meetings, visits, observations in meetings and presentations to fashion firms, and in-person conversations. In this way, expansion on the participants' previous responses provided multi-dimensional data which enriched the case studies. This utilisation of probing inquiry (Tsang, 2013) was made possible through the relationships that emerged through more extensive interactions with the 2 start-ups, and enabled the participants to introduce noteworthy matters that could have otherwise been overlooked.

Additionally, in-depth, semi-structured interviews were undertaken with 15 mid-sized and small-scale fashion brands that proffer sustainability as their leading unique selling point. The interviews with fashion firms provided qualitative insights into the specific challenges, needs, and perspectives of the fashion industry regarding supply chain and technological integration. Template analysis was deployed, allowing for the inclusion of *a priori* themes – themes which are predetermined before the coding process begins. By doing so, the research was able to concentrate on the three areas crucial to the study. As advised by Brooks et al. (2015), template analysis codes should be aligned to key theories found throughout the existing literature. In this study they were limited to the challenges of cost, scale, and data disclosure. Supplementary information was also collected from organisations' websites, white papers and third-party industry reports. Connecting the insights from the fashion firm interviews with the tech start-up cases helped to validate the relevance and effectiveness of the tech solutions, making a compelling prediction of tech adoption influences within the industry.

3. Findings

Case 1: Digital Product Passport solutions provider A

DPP solutions provider A (DPPA) is a technology firm that was established in Sweden in 2019. DPPA was selected for this study because of its effective utilisation of digital technologies on a large scale. The company uniquely operates both a technology enterprise and fashion manufacturing. The DPPA app offers comprehensive insights into fashion products, catering to many stakeholders throughout the supply chain. This objective is facilitated by the utilisation of a digital ecosystem, which is constructed on a publicly accessible blockchain platform (Ethereum). Prior to delving into DPPA's proposed resolutions for the existing issues encountered by SCT, it is necessary to acknowledge that DPPA approaches transparency through two sustainability dimensions, namely social performance, and environmental performance. The founder explains:

> In terms of environmental sustainability, consumers can view each step of the process [which is] based on the consumptions of the materials. The system calculates the amount of CO2 and water used to produce the product.
>
> Now the social aspect […] the app connects the people with the product. We want to make sure the technology is creating red lines, like no slavery. Our systems verify workers age, gender, and wage type […] and if it's above the minimum wage.

This observation highlights the organisation's emphasis on ideals rather than commercial forces in shaping its culture. When queried regarding the impetus for the organisation, the textiles engineer/CEO provided a comprehensive account of his own encounters with corruption, disparities, and fraudulent credentials prevalent inside fashion manufacturing facilities. The solution ensures that users can observe that the workers engaged in the production of each "DPPA'd" item receive equitable remuneration, and that every product is manufactured with minimal environmental consequences. The information presented is grounded in scientific principles and has been rigorously researched for accuracy and validity. The ownership of information is non-exclusive, as it is decentralised—a fundamental characteristic of blockchain technology.

DPPA also reported on their experience running an initial blockchain pilot with a Swedish fast fashion brand. The primary objective of the fashion brand's 'Openness Project' was to investigate the potential benefits of utilising automated methods for gathering sustainability data. Additionally, the project aimed to expand the limits of 'openness' (of data disclosure) within this context. The fashion brand also conducted a comprehensive analysis to determine the precise expenses associated with each garment, thus demonstrating economic transparency. This assessment shed light on the true production costs of garments within the Swedish context. The clothing items were produced within a global-local system that strives to maintain a high level of transparency. This allowed the implementation of a delayed assembly process, which resulted in a reduction in overproduction, and consequently a decrease in waste generation. The cotton used in the production of the fabric originates from Australia, while the textile manufacturing process takes place in Pakistan. The last stage of garment assembly by the production house occurs in close proximity to the brand's headquarters and distribution centre in Borås, Sweden. Each garment is equipped with a (smart phone readable) NFC chip, allowing customers to gain comprehensive visibility into the entire value chain, encompassing both social and environmental aspects, via the DPPA app. Consequently, consumers are able to access information about the origin of their clothing, establish trust in the sourcing process, be reassured of compliance with any existing responsible criteria set by the brand (and/or the regulators), and gain a comprehensive understanding of the intrinsic worth of the garments they have purchased. Both DPPA and the fashion brand were satisfied with the initial experiment and went on to develop a second project with greater numbers – thus trialing implementation at scale.

Case 2: DPP solutions provider B

Founder of DPPB's career initially focused on using digital technologies for education, skills training, microfinance, entrepreneurship, women's empowerment, ethical fashion, and fintech in developing economies. Prior to setting up DPPB, the CEO founded an ethical fashion label that produces sustainable and ethical products in Malawi. As the founder's career spans the digital, multimedia, non-profit, and retail sectors, it is no surprise that she has turned to technology in her focus on comprehensive global digital strategies that support transparency and ultimately the SDGs.

The strategy deployed by DPPB's technology arm centres around the development of digital identifiers for textile products. The Digital IDs serve as the foundation for an interconnected system that promotes circularity, sustainability, and regulatory compliance. The advantage of DPPB's strategy lies in its emphasis on interoperability, and the straightforward and secure interchange of information among various entities. Furthermore, their solution's accessibility to enterprises of all sizes is facilitated by the utilisation of a low code/no code framework. It is important to acknowledge that SMEs face considerable barriers in accessing various services within the industry. Consequently, DPPB's aim is to find ways to facilitate their inclusion – informed by their prior experience in this context. Their objective is to assist companies in establishing reliable connections with stakeholders through the use of trusted technologies like blockchain – in conjunction with dynamic QR codes, thus facilitating the communication of the fashion firm's circular credentials.

Additionally, DPPB aims to streamline the purchasing process for consumers by providing them with enhanced accessibility to second-hand products or rental services, thus supporting circularity. They do so by collaborating with a reseller start-up – thereby demonstrating their ability to develop a dual digital/circular ecosystem that will benefit numerous stakeholders. DPPB also offers integrated, bespoke marketing packages to SMEs. In this way they can work on the customer experience, for example embed the ID on their consumer facing website, or digitally connect a resell partner. Sometimes the required data is non-existent. At the end of the data collection phase a data map is created, and a gap analysis can take place and IDs can be populated. This can be automated, but also often results in copious manual work depending on data origins. Next, the information flow is tested and validated with the firm. Corrections, adjustments and unexpected tech challenges, plus user testing follows. Once tested, QR codes, product pages, and embedding on ecommerce sites can take place. However, the achievement of automated data input in an end-to-end chain of custody mechanism remains an aspirational goal.

In contrast to DPPA's implementation of solutions within a large off-shore manufacturer for a global brand, the investigations of DPPB for this study followed the onboarding of tech solutions by a small organic brand in the UK. The 10-year-old brand already has an established near-shored (in the EU) supply chain openly self-disclosed on its website. It has a loyal following of consumers – the majority of which are not local to the brand's headquartered location. The researcher attended meetings between the client and the tech start-up observing the questions, discussions and challenges that arose between the two parties (attended by the brand's proprietor and the DPPB's technicians). It was evident that both parties had values in common from the outset. However, the brand's spokesperson made several obstacles quite clear. This included the presumed high cost of the service, (although free for the pilot study), the cost of the (third party) resale service (also free for the pilot), and the time required for data input (if at scale). Of less concern was the 'fear of disclosure' of information to competitors – as the brand already shares information on their website. The research also found that the data points to be added were neither clear nor standardised – either from a fashion industry jargon standpoint and or legislative standpoint. Nonetheless, the brand's proprietor expressed a positive and collaborative mindset towards tech adoption and continued the discussion with DPPB for an expanded trial including a reseller solution.

4. Discussion

The theoretical contribution of this article lies in its analysis of the role of technology start-ups in driving advancements in SCT, and the role of blockchain technology in enhancing digital ecosystem thinking. This contribution highlights the paradox of dual tensions: the significance and the difficulty of 'collaborative' digital solutions. The study emphasises that technology alone is insufficient to drive SCT. Instead, it argues for synergy between technological adoption and socio-organisational levers within the fashion system. This holistic view represents a departure from more technologically deterministic approaches that see digital transformation as the answer – but overlook the importance of organisational culture and collaboration. This section provides evidence for this argument via the three *a priori* variables that were selected for this study: cost, scalability and data disclosure.

4.1 The Challenge of Cost Implications

Scholarly literature extensively discusses the economic considerations linked to the use of Web3 technologies that enhance supply chain transparency. These investigations have demonstrated that the primary deterrent to widespread acceptance is attributed to the substantial financial investments and complexities involved in achieving comprehensive worldwide traceability (Garcia-Torres et al., 2021; Kumar et al., 2017; Macchion et al., 2015; Boiral and Gendron, 2011). Yet fashion firms have other priorities – to ensure profitability by catering to client demands – and avoid 'unnecessary' costs (Gong et al., 2022). It became evident throughout the interviews for this study that the assumed financial implications associated with the implementation of the technology remain a significant obstacle in attaining widespread acceptance among enterprises.

However, the fear of litigious exposure is beginning to subvert brands' economic decision making. Beyond legislative compliance, brands fear that their perception in the marketplace may be tainted if they do not demonstrate verifiable evidence of sustainable practice. There exists a positive correlation between the level of transparency exhibited by a business and its financial performance, suggesting that increased transparency is likely to result in higher profitability (Fraser and van der Ven, 2022). Investors are also demanding equitable, environmental, social and governance (ESG) standards, hesitating to invest in firms with little evidence of corporate social responsibility (CSR). DPPA has successfully established a comprehensive digital ecosystem that offers four key dimensions: factory-facing, brand-facing, consumer-facing, and product-facing attributes. The cloud-based platform stores data on the assessment and compliance of factories in relation to ESG standards. This verification process enhances the attractiveness of factories to brands as well as investors. The utilisation of the DPPA ecosystem enables brands to increase the prices of their products by facilitating the mapping and management of their material assets and supply networks. Some evidence indicates that consumers are increasingly willing to pay a higher price for products that are environmentally sustainable (Dangelico et al., 2022). Indeed, this theory was confirmed in the fashion brand x DPPA pilot that saw the – blockchain enabled – higher priced collection sold out within three days of its initial launch.

It is hypothesised in this study that small scale firms are more likely to be successful in the digital transformation of the supply chain because of their sustainability values and motivations. These factors are keenly brought to bear in the pilot study conducted by DPPB with a small organic fashion brand in the UK. Engaging in the pilot with DPPB demonstrated the brand's openness, indeed eagerness to verify its claims. When asked why they would implement a blockchain-based technology over their (simpler) self-disclosure, they cited reassurance and the guarantee that their information was validated, safe and suitably compliant – no matter what the regulations. It is important to note that their willingness to experiment with the pilot in the first place came from their motivations and values – including pride in their sustainable supply chain and practice. That said, cost would still remain a barrier. Nonetheless, the brand would invest in the technology (given a reasonable price) as this would be one less task to deal with, and they could be assured of compliance with 'no trouble'. Indeed, lessening litigious exposure appears to be the biggest factor in moving brands to adopt and pay the accompanying fees for tech services. At this point there was little evidence that cost savings (in more efficient, risk-free) supply chains would motivate fashion firms to invest. This result will only become evident once firms adopt at scale, and collect data over time.

When queried regarding the cost to stakeholders, the founder spoke about DPPA's regenerative business model, that he referred to as a 'digital ecosystem'. The founder underscored the diverse value propositions that arise from this novel regenerative model for all three stakeholder categories, namely manufacturers, brands and consumers. The proposed regenerative business model by DPPA is founded upon the fundamental principles of incentivisation. The existing digital ecosystem of DPPA confers benefits upon (thus incentivising) all parties. This, in turn, motivates brands to allocate resources and enhance their present business models through innovation (thus 'regenerating' according to the CEO). This observation highlights the potential impact of a values-driven organisation. The DPPA case exemplifies a dual advantage for fashion brands adopting a regenerative business model. First, it showcases the potential of technology as an advantageous supplement, where the expenses of research and development are assimilated. Second, advantage can be attained through collaborative agreements within a resilient digital ecosystem, wherein participants offer reciprocal contributions towards the advancement of solutions (Rennie et al., 2019). This suggests that firms are inclined to deviate from their customary market strategies in order to seek mutually beneficial synergies. This reconsideration of organisational culture represents a shift in the fashion supply sector.

While regenerative business models buoyed by incentivisation may seem an emerging approach, the current business climate is focusing on more circular models. However, many companies lack adequate digitalisation to effectively engage in circular business models that yield financial benefits, thus potentially excluding such strategies from their strategic considerations. Insufficient knowledge regarding the post-sale fate of their products diminishes companies' inclination and capacity to actively participate in circular practices. Brands that lack the capacity to independently manage resale, rental, and repair activities, and lack the necessary level of digital proficiency to participate in established initiatives, are subject to lost financial benefits and opportunities. DPPB has recognised this potential new revenue stream for brands and therefore collaborates with a resale tech start-up to add value to the complete package they can offer to brands. By undertaking this additional task, it is addressing the need for individual IDs that are carried through both closed and open loop circularity systems. Yet, this gives rise to other difficulties such as automated data ingestion as well as interoperability – the ability for software programmes to interact with each other's programme languages and protocols. DPPB is tackling these challenges through the utilisation of advanced technologies such as artificial intelligence (AI), with a specific emphasis on the integration of low/no code solutions. This enhances the accessibility of transparency and circular economy models to under-resourced microenterprises and SMEs. Experimenting with their reseller and their brands, DPPB is not only seeking incentives but finding solutions that consumers will be able to deploy – for example using pre-loaded data to resell wardrobe discards.

Meanwhile, with the advent of advanced technologies, at the very least, sustainable practice can be communicated to the firm's advantage. For example, supplemental information that may be of interest to the consumer can be verified beyond former capabilities and thereby promote the firm in a more positive light, increasing brand reputation and eschewing unwarranted greenwashing (de Freitas Netto et al., 2020). Furthermore, the advantages to the SME of promoting a sustainable practice business model offers a more positive, multi-layered picture of openness and responsibility on the part of the firm (Todeschini et al., 2017). Through increased knowledge, improvements could be determined for sustainable SCT and ultimately systems change for the better.

4.2 The Challenge of Scalability

The cost implications bring us to the question of scale. The recognition of blockchain's potential is widespread. However, the challenge of scaling Web3 technologies is a complex matter that perplexes industry professionals, solutions architects, and scholars alike (Caldarelli et al., 2021). Some have referred to this challenge as 'blockchain's moon race' (Kenny, 2019), which indicates the competitive drive among start-ups to be the first to enter the market—or at least the first to be widely accepted. The promotion of transparent models necessitates the establishment of technological infrastructure and revised terms of collaboration. This is of particular significance, at the least because of looming and or enacted legislation, and at the most because of the urgent need to meet SDG targets. It is important to note that the EU Commission – the driving force behind the majority of compliance regulations – does not expect firms to invest in expensive technology. It suggests that all could be done manually on spreadsheets (Legardeur and Ospital, 2024; CIRPASS, 2023). However, this is untenable in the eyes of most firms – particularly at the granular level required. Identifying individual products poses a gargantuan task even for small firms. The Aura blockchain consortium claim to have already stored 40 million data points on fashion products (Bain, 2024). According to Schatsky, Arora and Dongre (2018), the primary obstacle to the adoption of blockchain technology has been its sluggish transaction processing speed, which restricts its effectiveness. As a result, many suggest blockchain is better deployed at the batch level – rather than the individual identifier. Based on the findings derived from the case studies, both DPPA and DPPB have effectively designed, tested and evaluated the application in collaboration with SMEs to perform at an individual item level – however these are still pilots with limited numbers. Furthermore, the system cannot be deployed at batch level – as each garment holds the potential of resale, and will have a unique product history. This begs the question why consortia are not sharing information to ease the data collection burden. For example, information could be shared on generic components such as zippers and buttons – which individually do not constitute a competitive advantage – but become very significant in the recycling phase of the product. At this point, no foresight is evident regarding such shared product information platforms. This data would ideally be stored using blockchain to ensure decentralised access.

Nonetheless, based on their pilots, DPPA has reached a significant stage of scalability. It is the more advanced solution of the two providers in this study. DPPA is now collaborating with a Levi's denim manufacturer in Pakistan:

> Surprisingly, our initial analysis revealed that the implementation of the application is significantly more feasible in larger supply chains compared to smaller ones. Consequently, we are presently engaged in the development of integration tools for companies' Enterprise Resource Planning (ERP) systems. This development aims to facilitate a seamless plug-and-play functionality, which aligns precisely with our ongoing efforts.

The collaboration between Levi's and DPPA serves as proof of the latter's capacity to offer a viable solution at scale. This directly challenges the obstacles identified by Schatsky et al. (2018), that is, technical feasibility and expansion of consortia. However, it is vital to acknowledge that this progress in scalability can only be achieved within a fashion supply chain that demonstrates a willingness to exchange data, possesses a power structure that is more equitable, and fosters a greater degree of collaboration compared to the prevailing standard (Heim et al., 2022). As mentioned earlier, DPPA has manufacturing connections with Levi's and therefore is in a position to leverage this arrangement, and shared strategic values.

Consideration of the effort expended by firms in the collection, 'cleaning', and retrieval of pertinent data is taken into account by DPPB's solution. DPPB sees data analysis as the heavy lifting in the digital transformation of the supply chain. Data sits across various company systems and departments, or with suppliers. The brand is dependent on the use of the tech provider's platform, to add information, create reports, update IDs and so on. This 'software-as-a-service' model is already quite commonplace, for example in enterprise and/or inventory management systems, such as product lifecycle management systems (PLMS), and accounting software systems. DPPB aims to integrate the systems as much as possible to facilitate interoperability and avoid duplication of effort—thus also saving business transaction costs. Altogether, the digital ecosystems behind effective data collection, storage and sharing for supply chain transparency has not been definitively resolved and remains a challenge for the tech start-up and its fashion clients. However, the desire for positive outcomes is shared across the stakeholders investigated for this study and is likely to drive speedier change.

The facilitation and achievement of SCT on a large scale is feasible but, at present, due to the ongoing development of the technology, it can only be accomplished by connecting with business partners who share similar goals and values. The fact that DPPA seeks engagement from 'forward thinking brands' indicates its awareness of its own organisational culture that shares similar values and perspectives. As seen in this study's findings, it is advisable for fashion companies to have a receptive stance towards the possibility of establishing partnerships and collaborations with technology companies, rather than regarding technology as a mere reactionary measure to address challenges related to SCT. The DPPA CEO additionally made reference to the existing deficiency of technology resources within fashion companies:

> Even prominent corporations often lack an internal technology department, and if they do own such a team, their focus is mostly on the development and maintenance of online retail platforms, typically at a rudimentary level. One advantageous aspect of our team is its composition, which consists of individuals specialising in three distinct fields: textile engineering, technology engineering, and research.

Therefore, technology is not the only barrier to scalability. The results of this study indicate that in order to effectively implement SCT on a large scale, it is also necessary to bring about cultural changes at the organisational level. This may be achieved by promoting the benefits of embracing a digital ecosystem mindset, wherein fashion and tech enterprises engage in joint efforts as if they were a single entity. This transformation entails shifting from a market-oriented culture to a values-oriented culture that prioritises innovation. Alternatively, fashion companies can establish flexible collaborations with organisations that embody these characteristics.

4.3 The Challenge of the Non-disclosure Paradox and Competitive Advantage

The data analysis also revealed that the absence of a globally recognised standard for blockchain deployment (Caldarelli et al., 2021) has resulted in increased competition among different technology companies that aspire to lead in traceability solutions. The escalation in competition has given rise to a novel occurrence referred to as 'tech-washing', which entails the act of affixing contemporary buzzwords onto pseudo-technological solutions. 'Tech-hushing' is also occurring when firms are actively using blockchain in the background but do not refer to the term blockchain because of its associations (and confusion) with cryptocurrency and disreputable dealings.

Much as tech companies are managing the disclosure (or not) of their competitive advantages, so too fashion firms are holding on to their 'trade secrets'. Confidential and sensitive information includes their sources and suppliers. The reluctance of fashion firms to disclose data is managed with DPPB's solution that offers varying degrees of information privacy with different parties, based on specific requirements. Digital identification systems facilitate the acquisition and dissemination of crucial product and material data, thereby enabling enterprises to actively engage in the circular economy. Additionally, this approach enables firms to obtain a comprehensive understanding of their environmental, social, and governance (ESG) performance and circularity status, including identification of potential hazards and areas for improvement.

The lack of understanding of the capabilities of tech solutions also suggests additional support is essential for SMEs. Support and guidance may be offered through insights from research and ensuing reports as well as the whitepapers published by tech companies. It may also be found through investment, affirmative government policy, mentoring, educational programs, constructive media/marketing stories, further innovations, as well as guidelines and frameworks on how to master these challenges (Doyle et al., 2022; Arthur, 2023). In the first instance, discovering, understanding, and then codifying strategies on digital transformation and technology adoption is likely to facilitate successful sustainable supply chain management and its digitalisation.

This study of blockchain technology as an enabler for positive impact in the fashion industry, its communities, and the planet is demonstrated through the case studies presented above. Both tech start-ups have emerged from somewhat diverse perspectives and are at different levels of maturity. Both have offered a window into the challenges faced by solutions providers and in doing so, answer some of the apprehensions around cost, scale and disclosure – of a technology that appears to be so promising. However, unlike the ease of use offered by infrastructures such as the Internet, social media and shopping apps, working with an advanced technology such as blockchain is still beyond the capabilities of most users. In its present structure it requires the intervention of specialists like the tech start-up cases here. They too struggle with certain aspects of the system's architecture, not to mention the nuances of the fashion supply chain. That said, as suggested by De Filippi and Lavayssière (2020) the democratising potential of the blockchain infrastructure is its hidden treasure, yet to be unlocked and shared for distributed power and decision making, and the universal good.

Conclusion

This study offers new perspectives on the integration of blockchain technology in fashion supply chains, highlighting the concurrent need for technological infrastructure and organisational transformation. It fills a gap in the existing literature by providing empirical insights and addressing specific barriers to advanced implementation of blockchain and associated technologies for the fashion supply chain. Aligning with the views of economics and enterprise theorists, blockchain technologies can enable economic benefits, risk aversion and more ethical business conduct (Davidsson et al., 2018; Berg et al., 2019). These factors affect almost every facet of the fashion supply chain globally.

However, the initial hypothesis of this thesis that the 'hard sell' of seeking the improved cost benefit ratio first, then the 'soft sell' of organisational change second, does not appear to have played out. Rather it is common values and desire to be seen as responsible (coupled with sustainability legislation) that motivates firms to experiment with the tech. The barriers of cost, scalability and disclosure are weighed up in perceived benefits of 'doing good'. The economic benefits this might bring are thus relegated to the future. Legislation has certainly accelerated interest in technology adoption and in a sense leveled the playing field – as all firms, large and small will be subject to these regulations. However, as solutions like those offered by DPPA and DPPB become more refined, other barriers remain that warrant further investigation. These may include longitudinal studies on the effects of incentivisation and return on investment to potentially offset technology costs. Additionally, effective collaboration, data sharing and interoperability are areas that remain open for investigation. Universally accessible solutions – enabled by the democratising capabilities of blockchain technology – that share information openly, validly and accurately will be increasingly useful in a potential future paradigm of shared and coordinated data for the common good.

References

Agrawal, T.K., Kumar, V., Pal, R., Wang, L. and Chen, Y. (2021). Blockchain-based framework for supply chain traceability: A case example of textile and clothing industry. *Computers & Industrial Engineering*, *154*, 107130. DOI: https://doi.org/10.1016/j.cie.2021.107130.

Amed, I., Balchandani, A., Marco, B., Berg, A., Hedrich, S. and Rölkens, F. (2019) *What radical transparency could mean for the fashion industry*. McKinsey. https://www.mckinsey.com/industries/retail/our-insights/what-radical-transparency-could-mean-for-the-fashion-industry.

Arthur, R. (2023). *The Sustainable Communication Playbook*. Nairobi: UNEP and UNFCCC. https://www.unep.org/interactives/sustainable-fashion-communication-playbook/?

Bain, M. (2024). Aura Blockchain Consortium Says It Has Logged More Than 40 Million Products. *Business of Fashion*. https://www.businessoffashion.com/news/technology/aura-blockchain-consortium-says-it-has-logged-more-than-40-million-products/ 2024] Review: Vol./issue. no., pg. no. missing.

Bak, G. (2019). Fashion, Blockchain, & AI, … Oh My! *The Startup*. https://medium.com/swlh/fashion-blockchain-ai-oh-my-afc161db3fdf 2020] Review: Vol./issue. no., pg. no. missing.

Behr, O. (2018). Fashion 4.0-Digital Innovation in the Fashion Industry. *Journal of Technology and Innovation Management*, *2*(1), 1–9. https://doi.org/10.1002/bse.701.

Benstead, A.V., Mwesiumo, D., Moradlou, H. and Boffelli, A. (2024). Entering the world behind the clothes that we wear: practical applications of blockchain technology. *Production Planning & Control*, *35*(9), 947–64. 10.1080/09537287.2022.2063173 (Accessed 2024/07/03). Review; doi address correct? accessed date needed?

Berg, C., Davidson, S. and Potts, J. (2019). *Understanding the blockchain economy: An introduction to institutional cryptoeconomics*. Edward Elgar Publishing.

Boiral, O. and Gendron, Y. (2011). Sustainable development and certification practices: Lessons learned and prospects. *Business Strategy and the Environment*, *20*(5), 331–47. https://doi.org/10.1002/bse.701.

Bounds, A. and Hancock, A. (2023). EU nominee for climate chief wants global fossil fuel taxes. *Finanacial Times*. https://www.ft.com/content/f3398a19-ea7d-460a-8a69-1813d53f4865.

Bullón Pérez, J. J., Queiruga-Dios, A., Gayoso Martínez, V., & Martín del Rey, Á. (2020). Traceability of Ready-to-Wear Clothing through Blockchain Technology. *Sustainability*, *12*(18), 7491. https://doi.org/10.3390/su12187491.

Caldarelli, G., Zardini, A. and Rossignoli, C. (2021). Blockchain adoption in the fashion sustainable supply chain: Pragmatically addressing barriers. *Journal of Organizational Change Management*, *34*(2), 507-524. https://doi.org/10.1108/JOCM-09-2020-0299.

Casciani, D., Chkanikova, O. and Pal, R. (2022). Exploring the nature of digital transformation in the fashion industry: opportunities for supply chains, business models, and sustainability-oriented innovations. *Sustainability: Science, Practice and Policy*, *18*(1). 773–95. DOI: https://doi.org/10.1080/15487733.2022.2125640.

Christopherson, S. and Lillie, N. (2005). Neither global nor standard: corporate strategies in the new era of labor standards. *Environment and Planning A*, *37*(11), 1919–38. https://doi.org/10.1068/a3789.

Chui, M., Issler, M., Roberts, R. and Yee, L. (2023). *Technology Trends Outlook 2023*. McKinsey. https://www.mckinsey.com/capabilities/mckinsey-digital/our-insights/the-top-trends-in-tech.

CIRPASS (2023). *FAQs*. EU: CIRPASS. https://cirpassproject.eu/faq/ (Accessed: 20/12/2023). Cole, R., Stevenson, M. and Aitken, J. (2019). Blockchain technology: implications for operations and supply chain management. *Supply chain management: An international journal*, *24*(4), 469–83. https://doi.org/10.1108/SCM-09-2018-0309.

Crowe, S., Cresswell, K., Robertson, A., Huby, G., Avery, A. and Sheikh, A. (2011). The case study approach. *BMC Medical Research Methodology*, *11*(1), 1–9. https://doi.org/10.1186/1471-2288-11-100.

Dangelico, R.M., Alvino, L. and Fraccascia, L. (2022). Investigating the antecedents of consumer behavioral intention for sustainable fashion products: Evidence from a large survey of Italian consumers. *Technological Forecasting and Social Change*, *185*, 122010. https://doi.org/10.1016/j.techfore.2022.122010.

Davidsson, P., Recker, J.C. and von Briel, F. (2018). External enablement of new venture creation: A framework. *Academy of Management Perspectives*. 10.5465/amp.2017.0163. Review; doi address correct? Vol. 34 issue. no.3, pg. 311–332.

De Filippi, P. and Lavayssière, X. (2020) 'Blockchain technology: Toward a decentralized governance of digital platforms?', in *The Great Awakening: New Modes of Life amidst Capitalist Ruins*. pp. 185–222 [Online]. https://doi.org/10.1353/book.80766.

de Freitas Netto, S.V., Sobral, M.F.F., Ribeiro, A.R.B. and Soares, G.R.d.L. (2020). Concepts and forms of greenwashing: A systematic review. *Environmental Sciences Europe*, *32*, 1–12. https://doi.org/10.1186/s12302-020-0300-3.

Destefanis, G., Marchesi, M., Ortu, M., Tonelli, R., Bracciali, A. and Hierons, R. 'Smart contracts vulnerabilities: a call for blockchain software engineering?'. *2018 International Workshop on Blockchain Oriented Software Engineering (IWBOSE)*: IEEE, 19–25. Review APA style guide for conference proceedings.

Doyle, M., Simpliciano, L. and Crowley, J. (2022). *The Traceability Playbook*. Sweden: TrusTrace. https://trustrace.com/traceability-playbook-fashion-supply-chains (Accessed: 27/07/22).

Edelman, G. (2021). The Father of Web3 Wants You to Trust Less. *Wired*. https://www.wired.com/story/web3-gavin-wood-interview/ 2024].

Elkington, J. (1998). *Cannibals with forks: the triple bottom line of 21st century business.* Gabriola Island, BC; Stony Creek, CT: New Society Publishers.

EllenMacArthurFoundation (2024). *Fashion and the circular economy – deep dive*. https://www.ellenmacarthurfoundation.org/fashion-and-the-circular-economy-deep-dive (Accessed: 27/07/24 2024).

EllenMcArthurFoundation (n.d.). *Shaping incentives to enable a circular, low-carbon economy*. https://ellenmacarthurfoundation.org/covid-19-shaping-incentives-to-enable-a-circular-low-carbon-economy (Accessed: 09/10/2023).

Fitzgerald, B. (2006a). The transformation of open source software. *MIS quarterly*, *30*, 587–98. https://doi.org/10.2307/25148740.

Fitzgerald, B. (2006) 'The transformation of open source software', *MIS quarterly*, 30(3), pp. 587–598. DOI: https://doi.org/10.2307/25148740.

Francisco, K. and Swanson, D. (2018). The supply chain has no clothes: Technology adoption of blockchain for supply chain transparency. *Logistics*, *2*(1), 2. https://doi.org/10.3390/logistics2010002.

Fraser, E. and van der Ven, H. (2022) 'Increasing Transparency in Global Supply Chains: The Case of the Fast Fashion Industry', *Sustainability*, 14(18), pp. 11520. DOI: https://doi.org/10.3390/su141811520.

Garcia-Torres, S., Rey-Garcia, M., Sáenz, J. and Seuring, S. (2021). Traceability and transparency for sustainable fashion-apparel supply chains. *Journal of Fashion Marketing and Management: An International Journal*, *26*(2), 344–64. https://doi.org/10.1108/JFMM-07-2020-0125.

Gartner (2019). *Interpreting technology hype*. Gartner Hype Cycle: Gartner. https://www.gartner.com/en/research/methodologies/gartner-hype-cycle (Accessed: 03/03/2020).

Geldard, R. and Ellerbeck, S. (2023). *How much progress is being made on the UN's Sustainable Development Goals?* World Economic Forum. https://www.weforum.org/agenda/2023/09/un-sustainable-development-goals-progress-report/ (Accessed: 10/10/2020).

Gong, L., Jiang, S. and Liang, X. (2022). Competing value framework-based culture transformation. *Journal of Business Research*, *145*, 853–63. https://doi.org/10.1016/j.jbusres.2022.03.019.

Heim, H., Chrimes, C. and Green, C. (2022). Digital hesitancy: Examining the organisational mindset required for the adoption of digitised textile supply chain transparency. *Blockchain Technologies in the Textile and Fashion Industry*, 47–80 [Online]. Version. https://link.springer.com/chapter/10.1007/978-981-19-6569-2_3#citeas Review.

Keniry, L. (2020). Equitable Pathways to 2100: Professional Sustainability Credentials. *Sustainability*, *12*(6). https://doi.org/10.3390/su12062328.

Kenny, L. (2019) 'The Blockchain Scalability Problem & the Race for Visa-Like Transaction Speed', *Medium*. Available at: https://towardsdatascience.com/the-blockchain-scalability-problem-the-race-for-visa-like-transaction-speed-5cce48f9d44.

Kumar, V., Koehl, L., Zeng, X. and Ekwall, D. (2017). Coded yarn based tag for tracking textile supply chain. *Journal of manufacturing systems*, *42*, 124–39. https://doi.org/10.1016/j.jmsy.2016.11.008

Lay, R. (2018). *Digital transformation – the ultimate challenge for the fashion industry.* Deloitte Consulting: Denver, CO, USA. https://www2.deloitte.com/ch/en/pages/consumer-industrial-products/articles/ultimate-challenge-fashion-industry-digital-age.html 2021].

Leanne Kemp (CEO Everledger) – Working Blockchain Platform for Diamonds | BIC18 (2018). Blockchain Innovation Conference. Review APA style guide for conference proceedings.

Legardeur, J. and Ospital, P. (2024). *Digital Product Passport in the textile sector*. Brussels: Panel for the Future of Science and Technology (Accessed: 06/07/2024). Review APA style guide for conference proceedings.

Ley, K. and Martínez-Pardo, C. (2020). *6 ways to drive funding to transform the fashion industry.* The World Economic Forum COVID Action Platform. Online: Word Economic Forum. https://www.weforum.org/agenda/2020/01/funding-circular-fashion-industry/ (Accessed: 10/10/2021).

Macchion, L., Moretto, A., Caniato, F., Caridi, M., Danese, P. and Vinelli, A. (2015). Production and supply network strategies within the fashion industry. *International Journal of Production Economics*, *163*, 173–88. https://doi.org/10.1016/j.ijpe.2014.09.006.

Mainelli, M. (2017). Blockchain will help us prove our identities in a digital world. *Havard Business Review,* HBR Insights Series. https://hbr.org/2017/03/blockchain-will-help-us-prove-our-identities-in-a-digital-world Review.

Maitland, I. (2003) 'The Great Non-Debate Over International Sweatshops', in Shaw, W.H. (ed.) *Ethics at work: basic readings in business ethics.* UK: Oxford University Press.

Mao, C., Qin, S.F. and Wright, D. 'Sketch-based virtual human modelling and animation'. *International Symposium on Smart Graphics*: Springer, 220–223. –> Review APA style guide for conference proceedings. Work not cited in-text.

Mellick, Z., Payne, A. and Buys, L. (2021). From fibre to fashion: Understanding the value of sustainability in global cotton textile and apparel value chains. *Sustainability*, *13*(22), 12681. https://doi.org/10.3390/su132212681.

Meraviglia, L. (2018). Technology and counterfeiting in the fashion industry: Friends or foes? *Business Horizons*, *61*(3), 467–75. https://doi.org/10.1016/j.bushor.2018.01.013.

Morlet, A., Opsomer, R., Herrmann, S., Balmond, L., Gillet, C. and Fuchs, L. (2017). *A new textiles economy: redesigning fashion's future:* Ellen MacArthur Foundation. Available at: http://www.ellenmacarthurfoundation.org/publications).

Natanelov, V., Cao, S., Foth, M. and Dulleck, U. (2022). Blockchain smart contracts for supply chain finance: Mapping the innovation potential in Australia-China beef supply chains. *Journal of Industrial Information Integration*, *30*, 100389. https://doi.org/10.1016/j.jii.2022.100389.

Niinimäki, K., Peters, G., Dahlbo, H., Perry, P., Rissanen, T. and Gwilt, A. (2020). The environmental price of fast fashion. *Nature Reviews Earth & Environment, 1*(4), 189–200.

O'Dair, M. (2019) 'Blockchain: The internet of value', in *Distributed creativity: how blockchain technology will transform the creative economy*. pp. 15-30 [Online]. Version. Available at: https://link.springer.com/chapter/10.1007/978-3-030-00190-2_2

Openlink (2018). Blockchain food traceability can revolutionize the industry. *Openlink*. at: https://www.openlink.com/en/insights/articles/blockchain-food-traceability-can-revolutionize-the-industry/ (Accessed: 10/09/2019).

Patel, J., Thomas, S. and Pore, N. (2018). *A Blockchain-Based Framework for Apparel & Footwear Supply Chain Traceability*. New Jersy. Available at: A Blockchain-Based Framework for Apparel & Footwear Supply Chain Traceability. Review format.

Powell, W., Cao, S., Foth, M., He, S., Turner-Morris, C. and Li, M. (2022) 'Revisiting trust in supply chains: How does blockchain redefine trust?', in *Blockchain driven supply chains and enterprise information systems.* pp. 21–42 [Online]. Version

PwC (2018). *Blockchain is here. What's your next move?* https://www.pwc.com/gx/en/issues/blockchain/blockchain-in-business.html (Accessed: 05/10/2019). Review format

Queiroz, M. M., Telles, R. and Bonilla, S. H. (2019) 'Blockchain and supply chain management integration: a systematic review of the literature', *Supply Chain Management: An International Journal*, 25(2), pp. 241–254 https://doi.org/10.1108/SCM-03-2018-0143.

Radocchia, S. (2018). Altering The Apparel Industry: How The Blockchain Is Changing Fashion. *Forbes*. https://www.forbes.com/sites/samantharadocchia/2018/06/27/altering-the-apparel-industry-how-the-blockchain-is-changing-fashion/#828cd8029fb6 (Accessed: 10/09/2019).

Remington, C. (2020). Report calls for sustainable tech investment. *Eco Textile News*. https://www.ecotextile.com/2020012125567/fashion-retail-news/investment-essential-to-scaling-sustainable-tech-report-says.html (Accessed: 04/01/2021).

Rennie, E., Potts, J. and Pochesneva, A. (2019) 'Blockchain and the Creative Industries: provocation paper', *Analysis & Policy Observatory*, Available at: RMIT Blockchain innovation hub. DOI: 10.25916/5dc8a108dc471.

Rusinek, M.J., Zhang, H. and Radziwill, N. (2018). Blockchain for a Traceable, Circular Textile Supply Chain: A Requirements Approach. *Software Quality Professional, 21*(1), 4–24. https://gateway.library.qut.edu.au/login?

Changing the Future of Fashion with Blockchain Technology, 2017. Directed by Sargeant, N. UK: YouTube.Review APA style guide for Youtube video citation.

Scaturro, S. (2008). Eco-tech fashion: rationalizing technology in sustainable fashion. *Fashion theory*, *12*(4), 469–88. https://doi.org/10.2752/175174108X346940.

Schatsky, D., Arora, A. and Dongre, A. (2018). Blockchain and the five vectors of progress. *Deloitte Insights*. https://www2.deloitte.com/content/dam/insights/us/articles/4600_Blockchain-five-vectors/DI_Blockchain-five-vectors.pdf (Accessed 09.08.22).

Stake, R.E. (1995). *The art of case study research.* Thousand Oaks CA: Sage.

Swan, M. (2015). *Blockchain: Blueprint for a new economy.* Sebastapol, USA: O'Reilly Media, Inc.

Tallontire, A. (2007). CSR and regulation: towards a framework for understanding private standards initiatives in the agri-food chain. *Third World Quarterly*, *28*(4), 775–91. https://doi.org/10.1080/01436590701336648.

Tapscott, D. and Tapscott, A. (2016). *Blockchain revolution: how the technology behind bitcoin is changing money, business, and the world.* Penguin.

Tsang, E.W. (2013). Case study methodology: Causal explanation, contextualization, and theorizing. *Journal of international management*, *19*(2), 195–202. https://doi.org/10.1016/j.intman.2012.08.004.

UN/CEFACT (2021) *Traceability and Transparency in the Textile and Leather Sector, Part 1: High-Level Process and Data Model:* United Nations Centre for Trade Facilitation and Electronic Business (UN/CEFACT). Available at: https://unece.org/sites/default/files/2021-03/BRS-Traceability-Transparency-TextileLeather-Part1-HLPDM_v1.pdf (Accessed: 05/07/2023).

UNECE 2023. UNECE blockchain pilot for cotton paves the way for more sustainable fashion. Review.

van Haaren-van Duijn, B., Bonnín Roca, J., Chen, A., Romme, A.G.L. and Weggeman, M. (2022). The dynamics of governing enterprise blockchain ecosystems. *Administrative Sciences*, *12*(3), 86. https://doi.org/10.3390/admsci12030086.

Volpicelli, G. (2018). Does blockchain offer hype or hope? *The Guardian*. https://www.theguardian.com/technology/2018/mar/10/blockchain-music-imogen-heap-provenance-finance-voting-amir-taaki 2020].

Von Hippel, E. (2005) 'Open source software projects as user innovation networks', in Feller, J., Fitzgerald, B., Hissam, S.A. and Lakhani, K.R. (eds.) *Perspectives on free and open source software,* pp. 267–278.

Warren, S., Treat, D. and Herzig, J. (2019). *Get the full picture: Assessing blockchain's business value:* Accenture30/09/19). Available at: https://www.accenture.com/au-en/insights/blockchain/wef-building-value (Accessed: 20/10/2020).

Ye, Y. and Lau, K. H. (2018) 'Designing a demand chain management framework under dynamic uncertainty: An exploratory study of the Chinese fashion apparel industry', *Asia Pacific Journal of Marketing and Logistics*, 30(1), pp. 198–234. DOI: https://doi.org/10.1108/APJML-03-2017-0042.

Yin, R.K. (2013). *Case study research: Design and methods.* Sage publications.

2 Next-gen Business Architecture with Blockchain in the Service Industry

Murat Kizildag[1*], Mahala Geronasso[2] and Marcos Medeiros[3]

1. Introduction

Blockchain technology has emerged as a transformative force in business world, offering innovative solutions to age-old technological challenges across various industries. At its core, blockchain is a decentralized and immutable ledger that securely records transactions across a network of computers. Unlike traditional centralized databases, blockchain operates on a peer-to-peer network, where transactions are validated by consensus among participants, ensuring transparency and trust without the need for intermediaries. Such new advancements and breakthroughs in the digital arena are also not far removed from service-oriented corporations. The service industry, which refers to a sector of the economy that provides services rather than tangible goods, plays a crucial role in modern economies, contributing significantly to employment and economic output. Thus embrace of blockchain technology in the service industry has the potential to redefine business operations and lead to the next generation of business architecture, due to blockchain's transformative capabilities, such as:

- Decentralization and Trust
- Enhanced Security
- Smart Contracts and Automation
- Improved Efficiency and Cost Savings
- Innovative Business Models
- Global Accessibility and Inclusivity
- Ownership and Privacy

This chapter specifically focuses on the next generational (next-gen) business practices and key operational areas that blockchain improves and scales up when adopted. Specifically, this chapter contributes and complements the existing body of knowledge by:

- exploring innovative governance and consensus models, optimizing smart contract functionality, promoting environmental sustainability, and navigating regulatory landscapes within the service industry framework.

[1,2] University of Central Florida
[3] University of Virginia
Email: mahala.geronasso@ucf.edu[2]; marcos@virginia.edu[3]
* Corresponding author: murat.kizildag@ucf.edu

- highlighting reduced administrative costs associated with traditional transaction systems in supply chain, asset management, investments, and shareholder transactions in the service industry (e.g., lower compliance costs through automated processes).
- addressing critical challenges such as scalability, interoperability, and security. We have developed novel solutions that significantly improve the efficiency, reliability, and applicability of blockchain networks in real-world business contexts.

The following section discusses the foundations, rationale and the ecosystem of blockchain mechanisms along with architectural use cases and next-gen business practices in the service industry. This chapter also brings forth arguments on enhancing business practices through on-chain solutions with a wide variety of outlook and sustainability, environment, ESG issues, and practices linked to blockchain technology in the service industry.

2. The Foundations and Rationale of Blockchain

As with other novel digital developments and advancements, it is critical to comprehend and discuss the core foundations, components, and functions of Blockchain technology, its underlying architecture for businesses, its transformational effects on the service industry, and related communities and stakeholders. Therefore, this chapter provides the structural framework of blockchain technology, and delves deeper into the exchange of blockchain's economic value, and facilitative roles of transcending several business transitions for all of the different branches of service industry, where stakeholder interactions and global movement heavily occur across the globe (Pal et al., 2021).

2.1 Core Premises & Ecosystem

Blockchain technology is the nest of incorruptible and immutable decentralized digital ledger (aka distributed ledger technology – DLT), where events/records in a chronological time order are individually clustered into blocks, and are established with decentralized peer-to-peer network (P2P) of nodes that are responsible for transaction processing, eventually creating the DLT (Kumar et al., 2023). Each node participates in the role of a client and a server to jointly provide and control data without any loss of information and/or data. DLTs can be seen as the participants and players in the chain's overall infrastructure. This decentralized digital environment provides a basis for the notarization of data, where every node on the chain is publicly verifiable, and information on the chain is concurrently available to all participating members (computers) in a blockchain platform.

Data blocks are cryptographically connected to the prior committed block by digital signatures and consensus within the network of computers. The on-chain data flow is reliable and secure due to the consensus algorithms that demonstrate how the network's nodes decide and process which transactions to accept or decline without any data breach to the other parties (Guru et al., 2023). Taken all together, via cryptographic proof, consensus algorithms, nodes and their application in the P2P networks within the DLT, each event and/or record is symbolized with a binary form with hashes, so-called "Merkle Tree", which is the logical formation and execution of blockchain transactions.

In Merkle tree, block headers in each block are cryptographically linked to the prior committed block by digital time stamps. Each on-chain transaction (hashes) has to go through a specific block, and each block records a unique Merkle root. The next step in the algorithm is that the parent nodes are put into pairs and their hash is stored one level up in the tree. Hence, Merkel tree can also be viewed as a per-block tree of all transactions included in the block, which are carried to the end-users (e.g., consumers, suppliers, vendors, etc.) with digital IDs with millions of consensus network computers in a decentralized and secure communication environment, without any 3rd party involvement.

Additionally, layer protocols in the blockchain domains work together with a consensus to channel the nodes for the endless repetition of immutable transactions (e.g., sales, contracts, exchange of funds, etc.). For instance, layer two is a third-party integration used in conjunction with layer one to enhance the number of nodes and, as a result, system throughput, which is measured in transactions per second (scaling phase).

Clients (users) connect with peers and share data via a client-server architecture where clients request data or content from application servers (*Hardware Infrastructure Layer*), a linked list is a collection of chained blocks with data pointers (the location of the variables) where each block is filled with data and pointers to the block (*Data Layer*), Nodes also carry out blockchain transactions in order to accomplish a common purpose (e.g., fund swaps) (*Network Layer*), making the arrangement of the proper sequence, and making sure every client or transaction is safe in the agreement (*Consensus Layer*). The application layer comprises the programs that end-users can connect and communicate via scripts, application programming interfaces (APIs) and user interfaces and frameworks, etc. (*Application and Presentation Layer*).

2.2 The Architectural Transformation

Blockchain technology, initially perceived as complex and intangible due to its decentralized nature and cryptographic underpinnings, has seen significant developments in recent years. These developments have demystified many aspects of blockchain, making it more accessible and understandable for businesses and industries worldwide. Architectural transformation via blockchain refers to the profound changes that blockchain technology can bring to various industries and systems by providing a more secure, transparent, digital, verified, and efficient way of managing data, assets, and business transactions (e.g., vendor relationship, insurance, etc.). Recent studies and implementations have demonstrated the practical applications and benefits of blockchain, particularly in enhancing transparency, security, and efficiency in various sectors. For instance, blockchain's role in improving supply chain transparency and financial transactions has been well-documented (Tapscott and Tapscott, 2016; Kshetri, 2017).

The core accepted notion in the business world is that blockchain, consisting of distributed ledger technology (DLT) and smart contracts, can provide strong foundations for both current earnings and new revenue streams by supporting new inter-system orchestrations and financial communications (e.g., profit and loss statements (P&L)). One of several reasons why companies (especially service-heavy companies such as hotels, retail, hospitality, etc.) are giving so much attention to blockchain is because of blockchain's perceived unique technological capability to provide all players and stakeholders (aka participants on chain). However, a global consensus is actually that companies, including those service-oriented ones, can and should determine whether they should invest in blockchain by focusing on specific use cases and their market position based on the core business operations and values. For short-term business perspectives, blockchain's strategic value in business transformation is mainly in cost reduction in the core business operations. Due to "no-middle-man" or "negotiating agency on a commission" in global business architecture with blockchain ecosystem, costs (e.g., overheads and operating costs) can be taken out of existing processes by removing intermediaries or the administrative effort of record keeping and transaction reconciliation enabling the creation of new revenue streams for blockchain-service providers.

The exact strategic value a business derives from blockchain depends on its specific use case and industry that eventually leads to next-gen business practices in a better format. For instance, certain industries' fundamental functions are inherently more suited to blockchain solutions capturing the greatest value (e.g., financial services and healthcare). Financial service companies operate to verify the financial information and transactions for their clients that can closely align with blockchain's core transformative impact on assets (e.g., retirement, tax, fund transfer, etc.). These solutions enabled by blockchain systems can be especially useful in cross-border payments and trade finance reducing the number of necessary intermediaries and are geographically agnostic. In the next section, we delve deeper into how blockchain can transform business operations within the domain of service-oriented companies by offering several different novel technological breakthroughs.

3. Enhancing Business Practices Through On-Chain Solutions

Blockchain technology can be highly beneficial for service industries by providing increased transparency, security, and efficiency (Yaqoob et al., 2022). Here are several ways in which blockchain can be useful for service industries in addition to the cost reduction:

- *Customer Loyalty Programs*: Companies can issue blockchain-based loyalty tokens, which customers can easily redeem and transfer, increasing customer engagement and loyalty.
- *Content Provenance and Digital Identity Verification*: Blockchain can help protect intellectual property and track the provenance of digital content. For hotel, accommodation, and airline businesses that require identity verification, blockchain can reduce costs associated with manual identity checks and the risk of identity fraud.
- *Enhanced Booking and Reservation Systems*: Blockchain can create a secure and transparent booking and reservation system, reducing the risk of double bookings and ensuring that travelers get accurate information about their reservations via smart contracts facilitation to eliminate disruptions.
- *Data Sharing and Inter-Collaboration*: Blockchain can facilitate secure data sharing and collaboration among travel industry participants, such as airlines, hotels, and travel agencies, improving operational efficiency and customer experiences, especially for group travelers with multiple reservations.

These short-term use cases for service-oriented companies offer tangible benefits, including cost reduction, process efficiency, enhanced trust, and improved customer experiences, making blockchain a valuable technology for businesses seeking quick wins and competitive advantages. While blockchain has significant potential, it is not a one-size-fits-all solution, and its adoption depends on diverse contexts.

3.1 Digital Asset Management on Chains

The digital revolution has created a new class of assets unlike their tangible and conventional (e.g., fiat money) counterparts. Digital assets reside in the intangible world of ones and zeros but hold tangible value for business transactions, operations, and commerce. As their prevalence grows, so does the challenge of their management, trade, and verification. Blockchain is, therefore, the technology set to transform and redefine the digital asset environment (Nakamoto, 2008; Tapscott and Tapscott, 2016).

Traditional digital asset management often leans on centralized authorities as trusted intermediaries, such as banks or specialized agencies. While they have their merits, central systems pose risks of single points of failure or control. Blockchain eschews this centralized approach. Instead, it relies on a decentralized network of participants, creating a more resilient and transparent ecosystem (Ozdemir et al., 2020; Swan, 2015). Trust is no longer bestowed in this realm but earned through transparent processes and shared verification.

As blockchain matures, its application in digital asset management will likely expand to several minor edges and corners of business operations. For instance, one could foresee scenarios where hotel loyalty points are tokenized, allowing patrons to trade or combine them in a decentralized market (Pazaitis et al., 2017). Alternatively, when luxury resorts offer exclusive digital collectibles as souvenirs, their authenticity and rarity are guaranteed by blockchain. Therefore, the intersection of blockchain and digital asset management redefines value, trust, and ownership in the digital age (Tapscott and Tapscott, 2017; Liu et al., 2022).

In inventory and asset management, blockchain technology presents a transformative paradigm. It enables real-time data tracking, eliminates redundancies, reduces stockouts, and minimizes excess inventory (Kshetri, 2017). Blockchain's transparency ensures that every item, whether an asset or inventory, is accounted for every stage of its lifecycle, reducing loss, theft, or misplacement (Kouhizadeh and Sarkis, 2018). Moreover, as the hospitality industry evolves, the way in which assets such as real estate or even lease contracts for tangible products are procured and managed is undergoing a revolution. Hotels, restaurants, and travel companies looking to expand or diversify their footprint can leverage blockchain to streamline their asset acquisitions (Marr, 2018). Via smart contracts on the blockchain, these entities can seamlessly purchase, lease, or even fractionally own physical spaces, PP&E (Property, Plant, and Equipment), and FF&E (Furniture, Fixtures, and Equipment) items. Not only does this ensure proof of the asset's history and the terms of the agreement, but it can also automate the flow of payments, dividends, or any returns on investments, reducing administrative overheads and potential human errors. These smart contracts can auto-execute when specific conditions are met, such as periodic lease payments or earnings distributions. Overhead costs diminish while trust between parties grows (Buterin, 2014).

3.2 Supply-Chain Transparency

Through a "permissioned blockchain," physical and virtual transactions can be chronologically recorded, ensuring data validation and distribution across interconnected entities enhancing transparency, and fostering trust across complex ecosystems. As its applications grow (e.g., risk prediction, data accuracy, etc.), blockchain amplifies the resilience of global supply chains in service industries such as airline companies (Queiroz et al., 2019). Furthermore, blockchain is now advanced enough to integrate with other emergent technologies such as the Internet of Things (IoT), smart contracts, and artificial intelligence (AI) to secure further and optimize the supply chain. In such environments, supply chain transparency concerns logistical efficiency, brand integrity, trust, and competitive differentiation (Kshetri, 2017). Thus, blockchain technology is disruptive that its applications bring transparency to the business fronts in the service industry (Kizildag et al., 2019; Saberi et al., 2019; Tugay and Erdogan, 2023).

Moreover, some of the service industry leaders have already identified the potential of blockchain for supply chain transparency with regulatory scrutiny. For instance, Starbucks' endeavors in tracking coffee beans from farm to cup to ensure fair trade and build a brand narrative of care, quality, and commitment. Such initiatives set industry benchmarks, pushing competitors to match and elevate their supply chain stories (Saberi et al., 2019).

The enhanced transparency enabled by blockchain can be further leveraged to improve the quality of the supply chain in several dimensions, including: product authentication, quality control & compliance and optimal inventory management. Blockchain can validate product authenticity and deter counterfeits by facilitating unalterable and secure documentation of a product's origin and movement throughout the supply chain. Each product can be attached with a unique blockchain identifier to improve product authenticity (Kamath, 2018). Blockchain can also facilitate product quality monitoring throughout the supply chain, allowing for quicker detection and elimination of flawed items, which minimizes waste and enhances consumer contentment (Swan, 2015). Furthermore, blockchain aids in monitoring inventory quantities and refining inventory management practices, decreasing inventory expenses, and enhancing supply chain performance with real-time data to better predict demand and supply by ensuring optimal stock levels preventing potential losses. For instance, self-executing contracts (e.g., smart contracts) can automate various supply chain processes, such as payments upon delivery with digital stores on the blockchain, ensuring they are tamper-proof and easily accessible by relevant parties (Queiroz et al., 2019).

3.3 Digital Privacy, Security, and Confidentiality

The service industry, being data-intensive, stands at the crossroads of vast opportunities and immense vulnerabilities. With its revolutionary features, blockchain promises a new paradigm in digital privacy and security (Zyskind et al., 2015). As service industries increasingly shift towards digital platforms, the volume of personal and transactional data they handle expands exponentially.

In the context of private and consortium blockchains, data privacy is a critical feature. Unlike public blockchains where data is fully transparent, private blockchains allow for restricted data access and control (Zyskind et al., 2015; Underwood, 2016). Blockchain's architecture ensures that once data is stored, it cannot be altered without consensus, guaranteeing data integrity and building trust among stakeholders. In private blockchains, cryptographic techniques such as encryption and digital signatures, along with decentralized identifiers (DIDs), enable users to control access to their data. Encryption ensures that only authorized parties can decrypt and access the data, while DIDs allow users to manage their identity and grant selective access permissions (Zyskind et al., 2015).

Unlike centralized databases, which are single points of failure and vulnerable to hacks, blockchain's decentralized nature spreads data across nodes, making unauthorized data breaches highly improbable. Thus, through cryptographic techniques and decentralized identifiers, blockchain gives users more control over their data, deciding who can access it and for what purpose, providing an added layer of security allowing users to engage with services without the fear of data exposure (Nakamoto, 2008). Blockchain also allows for "Selective Disclosure," a well-known concept in cryptography and blockchain technology, as described by Zyskind et al. (2015). This means users can choose to share only specific parts of their data, ensuring irrelevant personal details remain undisclosed. A customer could, for instance, verify their age at a hotel for age-restricted services without revealing their exact date of birth. Furthermore, blockchain facilitates direct interactions between parties, reducing intermediaries and associated security risks. Once a transaction is recorded on-chain, it becomes a permanent part of the ledger, reducing disputes (Swan, 2015).

Blockchain technology offers a robust solution by ensuring that data remains secure and confidential through its decentralized architecture and cryptographic protocols in the service industry. Each transaction is encrypted and linked to the previous one, ensuring data integrity and preventing unauthorized alterations (Nakamoto, 2008). In private blockchains, access to data is restricted, providing enhanced confidentiality and security for sensitive information (Zyskind et al., 2015). The service industry, especially sectors like travel and tourism, relies heavily on data to enhance customer experiences and optimize operations. Airlines, hotels, and travel agencies capture extensive data on traveler preferences, behaviors, and transactions (Kshetri, 2017). Blockchain's decentralized nature eliminates single points of failure, making it challenging for hackers to breach the system. Each participant in the blockchain network holds a copy of the ledger, and changes can only be made with consensus from the network. This decentralized approach ensures that data remains secure and tamper-proof (Underwood, 2016; Tapscott and Tapscott, 2016). Moreover, encryption transforms sensitive information into unreadable code that can only be decrypted by authorized parties. Digital signatures verify the authenticity and integrity of transactions, ensuring that data has not been altered (Narayanan et al., 2016). To sum up, blockchain might not be the a guaranteed solution for all data-related challenges, but it undeniably heralds a paradigm shift in how the industry can handle data confidently in the service industry (Kizildag et al., 2019).

3.4 Smart IDs, Contacts, and Verification

The integration of blockchain through the deployment of smart IDs, contracts, and verification systems, provides a promising and transformative answer. Numerous digital services rely on user identities. Blockchain introduces a revolutionary approach to identity verification. Instead of relying on centralized databases that are prone to breaches, blockchain-based smart IDs offer users a digital identity that's both secure and easily verifiable. This decentralized approach ensures that each ID is tamper-proof and universally recognizable without the need for a central governing authority. This enhances user privacy and streamlines many processes in the service sector (Baser et al., 2023; Dogru et al., 2018). In a hotel check-in process, for instance, with a smart ID system, guests could swiftly verify their identities through a digital platform, bypassing the traditional, time-consuming process. This could extend to digital room keys, enabling guests to access their rooms via a smartphone application, further integrating digital identity into the hotel experience (Tapscott and Tapscott, 2017).

Smart contracts are self-executing agreements with terms written into code. When pre-set conditions are met, the contract automatically executes, enhancing efficiency and reducing fraud. In the service industry, smart contracts could manage reservations, automatically allocating rooms and sending digital keys upon payment (Buterin, 2014). Blockchain's verification system is a core strength in this realm. Once authenticated by the network, every transaction is added to an immutable ledger. This ensures every transaction's authenticity and permanence, building users' trust. Consider a scenario where a guest disputes a charge from a hotel. With blockchain's immutable ledger, both the guest and the hotel can trace back to the exact transaction, verifying the validity of the charge. This ensures transparency and significantly reduces disputes and related overheads (Ozdemir et al., 2023).

3.5 Intellectual Property Rights

Intellectual Property (IP) is another complex and it sits at the heart of innovation, differentiation, and competitive advantage, especially in the ever-evolving service industry. With its vast opportunities, the digital age also brings a maze of challenges regarding IP protection. The rapid dissemination of information, ease of replication, and the blurred lines of digital boundaries have intensified threats to IP rights.

Blockchain emerges as a beacon of hope in this landscape by recording the entire lifecycle of an IP asset—from its inception to its various iterations and licensing agreements, all in a tamper-proof, transparent, and time-stamped manner. This immutable record-keeping property can deter IP infringements and facilitate smoother legal proceedings in case of disputes (Dogru et al., 2018). For instance, consider a travel tech startup that develops a groundbreaking algorithm for dynamic pricing. By recording its IP on a blockchain, the startup can create a concrete digital trail, making it more resilient against potential copycats or dispute with smart contracts. Thus, blockchain is not a simple technology; it is a paradigm shift in how the service industry can perceive, protect, and profit from its intellectual treasures; safeguarding IP rights becomes beneficial and imperative (Swan, 2015).

3.6 Stakeholder Relations

Navigating the intricate web of stakeholder relations is paramount in today's service industry, where reputation, loyalty, and trust can be as valuable as tangible assets. As businesses expand their digital footprint, the traditional dynamics of stakeholder engagement are being replaced by a realm that promises speed but is often mired in opacity, leading to trust deficits. This setting acts as a bridge, reconnecting businesses with their stakeholders through transparency, accountability, and automation (Treiblmaier and Beck, 2019). The days when customers, investors, or even employees had to take a company's word at face value are fading. They demand verifiable proof. Thus, Blockchain's inherent design ensures that every transaction, every claim, and every piece of data entered into it remains visible to those granted permission, setting the stage for an era of unmatched clarity and trust. Beyond that, blockchain offers a transparent lens into complex asset management aspects, such as compliance, insurance, fraud, and lease concerns. Traditional risk assessment models are outdated as blockchain's decentralized data analysis offers clarity in complex investments, such as deducing the true intrinsic value of hotel properties by evaluating some data points.

Another issue is that what if stakeholders had a way to trace every decision and every transaction back to its origin? Blockchain's precise nature facilitates the identification of errors in processing and lapses in service level agreements (SLAs), such as key risk indicators (KRIs). Therefore, stakeholders can better anticipate potential pitfalls, harness opportunities, and define their roles (Queiroz et al., 2019). Traditional contracts in the service industry, from supplier agreements to customer loyalty programs, often demand significant oversight to ensure compliance. Blockchain-based smart contracts promise to disrupt this. These automated processes, governed by smart contracts, assure stakeholders that terms will be executed impartially, fostering trust and reducing administrative burdens. For instance, during real estate endeavors, such as resort developments, shareholders are shielded by blockchain-drafted clauses allowing equitable buy-back or sale of shares. They are also guaranteed clarity in wealth management terms, whether it concerns dividend payouts or the returns on their investments (Yucel et al., 2023; Tapscott and Tapscott, 2017).

To provide a clear visual representation of the various application areas of blockchain within the service industry, we have created the mapping diagram below (Figure 2.1). This diagram illustrates the central role of blockchain technology in integrating and enhancing various aspects of the service industry, demonstrating how improvements in one area can positively impact others. The interconnectedness of these applications highlights the comprehensive potential of blockchain to revolutionize the service sector.

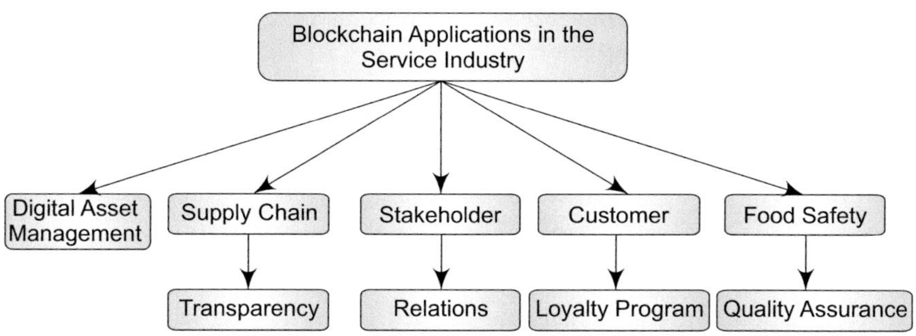

Fig. 2.1 Mapping Diagram of Blockchain Applications in the Service Industry

(*Source:* authors' creation)

4. Harnessing ESG and Sustainability with Blockchain in the Service Industry

Building on the previous discussions, the third part of this chapter focuses on the intersection of blockchain technology and ESG (Environmental, Social, and Governance) initiatives. By leveraging blockchain's inherent features of transparency, immutability, and decentralization, service-oriented companies can revolutionize their approach to sustainability, social responsibility, and governance. This exploration not only underscores current applications but also envisions how the service industry embraces blockchain technology to create changes in the realms of environmental, social and governance (ESG) (Grover et al., 2019).

Traditional social responsibility under ESG often involves businesses, like hotels, pledging a portion of their revenue to local social causes, with stakeholders largely taking these promises at face value or relying on third-party audits for validation (such as balance sheets or income statements). However, it is evident that blockchain technology is fundamentally reshaping the landscape of social responsibility within the service sector (Bedin et al., 2020). Its impact can be observed across domains such as hospitality management, tourism, energy, and the environment. This can largely be attributed to blockchain's ability to offer a transparent and decentralized platform for validating and overseeing social activities while emphasizing social and corporate responsibility (Liu et al., 2023). With blockchain's transparent and immutable ledger, a hotel could use smart contracts to automatically allocate a percentage of every room booking directly to a dedicated cause, such as local educaiton under one of the corporate social governance. Rather than merely stating intentions, the hotel would be demonstrably committing to its social pledges, allowing guests and investors to see the tangible impact of their contributions their contributions to the society.

With the help of blockchain, customers are now empowered to measure the impact of their purchases promoting sustainable commerce fostering transparent governance practices in the service industry. There has been a shift in customer preferences towards organizations that prioritize ethics and environmental responsibility, as the credibility of service businesses and corporations have noticeably improved due to the traceability and accountability offered by blockchain technology such as artificial intelligence (Wang et al., 2021).

4.1 Legal Aspects and Ethical Practices

The legal realm forces a reexamination of fundamental legal ideas such as the identification of legal people, property rights, and the nature of contracts. For instance, several foundations of the macroeconomic system, such as the reliability of currency and the function of central banks, are now being investigated. The experimentation with blockchain has resulted in a reshaping of microeconomic factors, such as asset scarcity and the ever-changing definitions of status commodities. When integrated into the service industry, a thorough examination of existing legal frameworks is required to protect established norms and stakeholder rights. Blockchain's immutable and transparent nature can bring about changes in resource distribution eliminating inefficiencies and ensuring fairness.

As the service industry embraces ESG considerations and integrates blockchain technology, the importance of ethical principles becomes even more apparent. The goal is to enhance trust, promote conduct and foster accountability. To ensure its impact and prevent any negative consequences, it is crucial to establish clear guidelines and standards for its use. These guidelines serve as a guiding light ensuring that the technology is applied in an ethical manner, especially in the service sector. This is particularly important as industries strive to find a balance between innovation and ethical considerations (Sharif and Ghodoosi, 2022). The decentralized nature of this fosters transparency and encourages morally upright corporate practices with a commitment to societal betterment. Furthermore, this technological advancement has prompted the service sector to embrace friendly and sustainable business approaches (Ronaghi and Mosakhani, 2021). Transparency, protection of information, fairness preservation, equal opportunity promotion, fairness maintenance, inclusion promotion and accountability acceptance are crucial factors to consider.

Taken all together, the fundamental aspects of human identification, governance, and the verification of information are all going through transformations on the social front. Concepts like sovereign identity, pseudosymmetry, and creative governance models are all fostered by blockchain technology, all of which pose a threat to conventional standards. The groundbreaking capability of technology to validate data without digging into its context has the potential to alter verification procedures across industries, including the service sector.

4.2 Trust and Social Engagement

The adoption of blockchain technology can help service related businesses to improve standards for social and environmental requirements. The growing focus on ESG in the service sector is driven by both requirements and commercial motivations. Within the realm of sustainability, blockchain technology has applications such as supply chain management and green manufacturing. It offers the potential for improved environmental and social performance across supply networks. Moreover, it excels at enhancing traceability and transparency in supply chains advocating for sustainable operations. The main goal here is to encourage friendly business practices while reducing ecological impact.

Trust plays a role in the service industry and can greatly benefit from blockchain technology. Its decentralized nature combined with record keeping ensures reliable transactional history thereby enhancing system trust while reducing reliance on intermediaries. For example, in the realm of copyright services, blockchain technology can enhance the reliability and accessibility of copyright data, speed up royalty payments and improve transparency throughout the value chain. It builds a sense of trust among users, creators, and industry leaders. Further, this technology empowers customers to seamlessly connect with service providers, provide feedback and influence product or service improvements. A notable example is its role in establishing a system for charitable donations during challenging times. Blockchain has the potential to increase transparency and credibility in the philanthropic sector by creating platforms where individuals have control over their own data. This bestows them with decision-making power and fosters a sense of ownership and active involvement.

4.3 Behavioral Dimensions

The potential effect and use of blockchain lies in its decentralized and transparent character, which can influence both behavior and decision-making processes for many stakeholders (Hernandez et al., 2023). It is important to highlight that the relationship between corporate behavioral dimensions and the disclosure of sustainable strategies (the degree to which a company reveals information on environmental, social, and governance aspects) can be impacted by many internal factors including the firm's size, industry, and liquidity (Modugu, 2020). Ultimately, the service sector is undergoing a transformation due to the use of blockchain technology. The use of blockchain technology has the potential to enhance endeavors aimed at enhancing social responsibility, optimizing supply chain management to foster environmental sustainability, facilitating streamlined governance processes, and influencing consumer behavior.

The capabilities of blockchain have a lot of potential to benefit stakeholders in many venues that make up the service sector, including hotels, restaurants, and retail shops. Consider, for example, the aspect of corporate governance that is concerned with trust and openness. The immutability of blockchain may give an auditable record of choices taken, which can be helpful for selecting board members for a hotel chain. In a similar vein, voting rights for decision-making processes may be safely and transparently conducted on blockchain platforms in the context of large corporations including restaurants chains and retail malls. This ensures that each stakeholder's vote, assets, and investments protected and tamper-proof in real time, thereby benefiting from an increased degree of both security and transparency. This builds an atmosphere that is transparent, equitable, and conducive to progressive development.

5. Challenges

Blockchain technology is on the brink of revolutionizing the service industry as it is for the other sectors all around the world. However, transforming this potential technology into real-world benefits presents many challenges. Challenges in implementing blockchain technology in the service industry can be. These include scalability dilemmas, market-related risks, uncertainties in transactions, and technological vulnerabilities. Despite the numerous added-value that blockchain landscape brings to the entire service industry such as guest management and cryptocurrency payments, direct bookings, reservations, and digital check-ins/outs, addressing these challenges is crucial for its wider adoption and future economic growth in the industry as a whole. This section delves into the present obstacles that blockchain technology faces.

5.1 Organizational Challenges

Implementing blockchain technology requires significant changes in organizational processes and structures. Many organizations face resistance to change, especially when it involves adopting new and complex technologies. Effective change management strategies are essential to facilitate this transition. Hotels, which have long relied on conventional reservation systems, will confront the significant challenge of incorporating this emerging technology. This integration will necessitate financial adjustments, operational shifts, and extensive staff training, transitioning from paper-based transactions to paperless digital disclosures (Peters and Panayi, 2016). Additionally, the seasonal fluctuations in the service sector contribute to the slow adoption of blockchain and other emerging technologies such as machine learning (ML) and artificial intelligence (AI) (Kizildag et al., 2019).

5.2 Technological Challenges

Blockchain technology, while promising, is still in its early stages and faces several technological hurdles. Scalability is a significant issue, as current public blockchain architectures could strain under high transaction frequencies, leading to system lags and increased costs in the service sector (e.g., travel bookings, multiple reservation arrangements for flights) (Zheng et al., 2017). Blockchain's need for uninterrupted internet connectivity further adds to the challenges, especially in remote locations where internet infrastructure may be sparse. Additionally, merging multiple blockchain systems might intensify scalability issues, considering their transactional pace (Dadkhah et al., 2022). Scalability emerges as a predominant concern, marked by slow transaction speeds and data duplication issues. Transaction uncertainties arise from irrevocable errors and the potential to cancel transactions, while technological challenges encompass difficulties in updating blockchain technology systems and the need to protect blockchain keys, (Biswas and Gupta, 2019). The current technological landscape will most likely be resolved through real businesses conducting real testing, usage, and consistent adoption over the mid-to-long term (Alomari and Fetais, 2023).

5.3 Data Confidentiality and Privacy Challenges

The service industry currently handles sensitive customer data in additional ways, which data confidentiality and privacy is not fully protected. Blockchain's immutability means that once data is added, it cannot be altered or deleted, posing significant privacy protection. For example, a luxury spa resort that collects detailed health and wellness data from its customers to personalize treatments must consider who has access to this data, how it is shared across the chain, and the potential for irreversible errors. The decentralized nature of blockchain can make data rectification complex (Zyskind et al., 2015).

5.4 Regulatory and Environmental Challenges

The regulatory landscape for blockchain technology is still evolving, adding another layer of complexity to reliable business transactions. Requirements for data protection and privacy vary across different jurisdictions, which can be particularly challenging for global service providers. Navigating these regulatory landscapes, especially when dealing with personal and financial data in booking and travel arrangements, brings ethical and governance issues. Decentralized autonomous organizations (DAOs) and blockchain-based governance models can introduce challenges in decision-making and conflict resolution (Finck, 2018). Additionally, blockchain networks, especially those using proof-of-work consensus mechanisms, face criticism for their high energy consumption. Achieving a balance between security and environmental sustainability remains a challenge (Kwok and Koh, 2019).

5.5 Cost-Related Challenges

The initial costs of implementing blockchain technology can be high, including expenses related to infrastructure setup, software development, and integration with existing systems. Ongoing operational costs such as energy consumption, network maintenance, and transaction fees must also be considered. Businesses must carefully evaluate the potential return on investment and ensure that the benefits of blockchain adoption outweigh the costs (Tapscott and Tapscott, 2016). Unlike their counterparts in banking and FinTech, major travel platforms such as Kayak and Expedia have slowly adopted and embedded digital innovations (Kizildag et al., 2019).

5.6 Environmental, Governance, and Social Challenges

Another layer of complexity to reliable business transactions, especially for governance (e.g., ESG, sustainability, trust, transparency, stakeholder conflicts, etc.). Businesses often find themselves charting a murky regulatory landscape, especially for services with global outreach as present in the global service sector. Different jurisdictions have varying requirements for data protection and privacy. Decentralized autonomous organizations (DAOs) and blockchain-based governance models can introduce challenges in decision-making and conflict resolution such as ethical dilemmas and disputes. In addition, energy considerations must be considered. Mining and validating operations on blockchain demand significant energy, although newer versions offer more energy-efficient alternatives (Kwok and Koh, 2019). In other words, proof-of-work blockchains like Bitcoin have faced criticism for their energy consumption. Achieving a balance between security and environmental sustainability remains a challenge for blockchain networks.

6. Coming up Next

Blockchain architecture and ecosystem is identified as a disruptive innovation landscape that combines tech attributes such as distributed data storage, point-to-point transmission, secured transactions, consensus mechanisms, and encryption algorithms. These are already and readily implemented in the so called "FinTech" industry as compared to other industries. This is one of the key reasons why other industries such as hospitality and tourism have not capitalized on the applications of this innovation, mainly due to its customer-centric business culture. However, in the foreseeable future, as the service industry seeks to harness the transformative potential of blockchain, there will likely be a convergence of traditional operational models with cutting-edge digital solutions.

6.1 Next-Gen Business Architecture with Blockchain Integration

Blockchain technology is poised to transform the next-generation business architecture by integrating with other advanced technologies such as artificial intelligence (AI), machine learning (ML), and the Internet of Things (IoT). These integrations will enable more efficient, transparent, and secure operations within the service industry. For example, combining blockchain with IoT can enhance real-time tracking and data verification in supply chains, ensuring that every item is monitored from production to delivery. AI and ML can optimize blockchain-based processes by predicting trends, automating decision-making, and enhancing data analysis (Tapscott and Tapscott, 2016; Mougayar, 2016). Additionally, smart contracts can be used to automate various business processes, reducing the need for intermediaries and minimizing human errors. In the foreseeable future, as the service industry seeks to harness the transformative potential of blockchain, there will likely be a convergence of traditional operational models with cutting-edge digital solutions. Hotels, airlines, and other service entities might establish hybrid models—where traditional systems coexist with blockchain until a smoother transition is feasible. This transitional phase will be marked by a steep learning curve, necessitating significant investments in training, infrastructure, and cybersecurity (Dadkhah et al., 2022).

6.2 Tackling Challenges and Considerations

Addressing the challenges associated with blockchain implementation is crucial for its widespread adoption. Scalability, regulatory compliance, and data privacy are some of the key issues that need to be tackled. Solutions such as layer-2 scaling techniques, which offload transactions from the main blockchain to secondary layers, can significantly improve transaction speeds and reduce costs. Regulatory sandboxes can provide a controlled environment for businesses to test blockchain solutions without fully complying with existing regulations, allowing for innovation while ensuring regulatory oversight. Privacy-enhancing technologies, such as zero-knowledge proofs and secure multi-party computation, can help maintain data privacy while allowing blockchain to be used for sensitive information (Finck, 2018; Zheng et al., 2017). Collaborative efforts among industry stakeholders, regulators, and technology providers will be essential in addressing these challenges. For example, establishing industry consortia can help standardize blockchain practices and promote interoperability between different blockchain platforms.

6.3 Promising Applications for Better Business Practices

Beyond the current applications, blockchain holds promise for several innovative uses in the service industry. For instance, blockchain can facilitate decentralized loyalty programs where customers earn and redeem points across multiple businesses, providing greater flexibility and value. Cross-border payments can be streamlined using blockchain, reducing the time and cost associated with traditional banking methods. Secure digital identities can be created on blockchain, allowing customers to verify their identity once and use it across multiple services, enhancing convenience and security (Kshetri, 2017; Narayanan et al., 2016).

Hotels might leverage blockchain for tamper-proof guest reviews, creating a new benchmark for authenticity. Airlines could employ blockchain to create a unified global passenger identity system, vastly simplifying international travel. Additionally, blockchain-powered loyalty programs could seamlessly intertwine with traditional reservation infrastructures, empowering customers with real-time loyalty point accrual and redemption, fostering heightened guest satisfaction (Jain et al., 2023). Blockchain technology establishes asymmetric encryption platforms, where lodging companies can issue loyalty tokens as rewards in their loyalty programs (Crosby et al., 2016). Service-oriented businesses that adapt and leverage blockchain's capabilities are positioned to benefit from upcoming developments. For instance, the tokenization of real-world assets, ranging from real estate, land, property, and equipment to commodities, will continue to gain momentum, making it easier to trade, transfer, and fractionalize ownership of traditionally illiquid and hard assets, potentially opening up new investment opportunities.

6.4 Widespread Adoption Beyond Small-Scale Experimentations

The transition from small-scale experimentations to widespread adoption of blockchain technology in the service industry requires robust pilot projects, real-world testing, and continuous improvements. Successful case studies and pilot projects can demonstrate the viability and benefits of blockchain, encouraging more businesses to adopt the technology. Additionally, industry standards and best practices will play a critical role in guiding the large-scale implementation of blockchain solutions (Tapscott and Tapscott, 2016). Developing comprehensive training programs and resources will be essential to equip industry professionals with the necessary skills and knowledge. Partnerships between academia, industry, and government can facilitate research and development, ensuring that blockchain technology continues to evolve and meet the needs of the service industry. It's also important to establish clear metrics and benchmarks to evaluate the performance and impact of blockchain implementations, ensuring that they deliver the intended benefits and continuously improve.

6.5 Enhancing Customer Experience and Operational Efficiency

Looking forward, the digitalization movement in the service industry is rapidly evolving, with blockchain technology at its core, further enhanced by generative artificial intelligence (GAI) in open-source spaces. Service-related businesses that quickly embrace these transitions will gain competitive advantages in customer management, guest experience, and service excellence due to early adoption benefits and potential network effects.

For instance, no "invasion of privacy" concerns for the guests will arise as they will be able to determine the level of information they would like to share via AI-enabled apps or portals for their personalized booking and guest services, providing better and more trustworthy customer service. Blockchain's trust-centric, transparent, and decentralized nature aligns more with industry needs, leading to broader adoption that transcends simple transactional uses. Moreover, businesses must be quick to adapt and innovate to keep up with the always-changing needs of their clientele in the entire supply-chain scheme. Transparency will increase in areas like investments, governance, and sustainability, lowering transaction costs and improving communication about service-related companies' strategic operations and financial plans and projections across all participating stakeholders. Thus, both service and hospitality sectors will transition toward more transparent, efficient, and value-driven, ethical operations (Ozdemir et al., 2023).

Conclusions

Blockchain technology introduces a new paradigm of trust, transparency, and efficiency, which has the potential to drive a significant transformation in service-oriented businesses. This chapter provided a comprehensive analysis of how blockchain can drive innovation in the service industry by enhancing operational processes, improving supply chain management, protecting digital identities, and revolutionizing financial transactions.

Summary of Key Findings

As previously discussed in detail, the foundational elements of blockchain technology include its decentralized architecture, cryptographic security, and consensus mechanisms, and these features can be leveraged to improve business operations in the service industry. For instance, businesses can enhance supply chain management by tracking products from origin to end-user, ensuring product authenticity and quality. Additionally, blockchain empowers individuals to have greater control over their digital identities, protects intellectual property, and revolutionizes financial transactions, cross-border payments, and the tokenization of assets. As blockchain continues to mature and find applications across industries, it is poised to reshape traditional business models, optimizing operations, and fostering innovation.

The importance of understanding the core mechanisms of blockchain spaces, technology, and its implementation to the main corporate functions have become more and more pronounced for institutions in divergent economies, especially in recent years. This is mostly because the rise of the digital customer, trust, and changes to the security landscape have resulted in intelligent automation (e.g., virtual payments), and the dominance of digital platforms in recent years (World Economic Forum, 2017).

Specific to the service industries, several promising applications were highlighted, such as decentralized loyalty programs, secure digital identities, and tamper-proof guest reviews, demonstrating blockchain's potential to enhance customer experiences and operational efficiency. These issues are mainly seen in service-oriented industries such as, travel, hospitality, transportation, etc., since such industries must deal with data breaches, consumer data protection, ESG issues, ethical concerns, stakeholder relationships, consumer confidentiality, etc. Given these vulnerabilities, mapping all the implementation areas of blockchain platforms and framing how allocating these innovations and features (e.g., transaction immutability, resiliency against cyber-attacks and fault tolerance) can revolutionize and optimize overall operational efficiency in the service industry.

Limitations

Despite its potential, the widespread adoption of blockchain in the service industry faces significant challenges. Scalability remains a critical issue, with current blockchain architectures struggling to handle high transaction volumes efficiently. Regulatory compliance and data privacy are also major concerns, as different jurisdictions impose varying requirements that can complicate implementation. Additionally, the high costs associated with integrating blockchain technology, including infrastructure setup and ongoing operational expenses, pose a barrier to entry for many service providers.

Future Work

While blockchain technology offers significant opportunities for the service industry, its successful implementation will require overcoming substantial challenges. Future research should focus on developing scalable blockchain solutions that can support the high transaction volumes typical in the service industry. There is also a need for standardized regulatory frameworks that can provide clear guidelines for blockchain adoption across different regions. Further exploration into privacy-enhancing technologies and their integration with blockchain will be crucial to address data confidentiality concerns. Pilot projects and case studies are essential to demonstrate the real-world benefits of blockchain and encourage broader adoption. Furthermore, the adoption and use of blockchain technology are not solely determined by its technical capabilities. Behavioral dimensions play a crucial role in how individuals, organizations, and societies interact with and make decisions regarding blockchain. For instance, trust is a fundamental behavioral aspect of blockchain. Users need to trust the technology, its participants, and the rules governing the blockchain network. Perceived trustworthiness is influenced by factors such as the technology's security features, transparency, and the reputation of network participants. Many blockchain networks use tokens and incentives to encourage participation and specific behaviors. The design of these incentive structures can influence how participants engage with the network. Behavioral economics principles, such as rewards and penalties, play a role in shaping behavior in the corporate world.

The impending decade heralds a phase of accelerated digital transformation via Blockchain implementation for the service industry. With blockchain poised to play a pivotal role, service entities have an unprecedented opportunity to meld traditional operational wisdom for their core business culture with blockchain's unparalleled cost-efficiency, transparency, trust, and communication. The industry can navigate this transformative journey through strategic investments and collaborative endeavors, leading to an era an era of enhanced operational efficacy and elevated customer experiences, travel norms, accommodation needs, and guest services (Kizildag et al., 2019).

Future research should focus on developing scalable blockchain solutions that can handle high transaction volumes without compromising efficiency. There is also a need for standardized regulatory frameworks that provide clear guidelines for blockchain implementation while ensuring compliance with international laws (Yaqoob et al., 2022). Exploring the integration of blockchain with other emerging technologies such as artificial intelligence, Internet of Things (IoT), and quantum computing could unlock new capabilities and enhance the overall efficiency and security of blockchain networks. Moreover, research into advanced privacy-preserving techniques, such as zero-knowledge proofs and homomorphic encryption, could help balance transparency with data privacy (Zhang et al., 2019). Collaborative efforts between academia, industry, and regulatory bodies will be essential to address these challenges and drive the adoption of blockchain technology in the service industry. By focusing on these areas, future work can ensure that blockchain technology not only meets current needs but also evolves to support the dynamic requirements of the future digital economy.

References

Alomari, A. and Fetais, N. (2023). *Blockchain Technology Adoption in the State of Qatar: Qualitative Risk Analysis*. The International Conference on Civil Infrastructure and Construction.

Başer, M.Y., Büyükbeşe, T. and Kizildag, M. (2023). What if we could travel without passport? First sight to blockchain-based identity management in tourism. *Asia Pacific Journal of Tourism Research, 28*(4), 341–63.

Bedin, A., Queiroz, W., Capretz, M. and Mir, S. (2020). A blockchain approach to social responsibility. 15–29. https://doi.org/10.1007/978-3-030-50072-6_2.

Bertino, E., Kundu, A. and Sura, Z. (2019). Data transparency with blockchain and AI ethics. *Journal of Data and Information Quality (JDIQ)*, *11*(4), 1–8.

Biswas, B. and Gupta, R. (2019). Analysis of barriers to implement blockchain in industry and service sectors. *Computers & Industrial Engineering*, *136*, 225–41.

Buterin, V. (2014). A next-generation smart contract and decentralized application platform. white paper, 3(37), 2–1.

Boutkhoum, O., Hanine, M., Nabil, M., Barakaz, F., Lee, E., Rustam, F., … and Ashraf, I. (2021). Analysis and evaluation of barriers influencing blockchain implementation in moroccan sustainable supply chain management: an integrated ifahp-dematel framework. *Mathematics*, 9(14), 1601. https://doi.org/10.3390/math9141601.

Crosby, M., Pattanayak, P., Verma, S. and Kalyanaraman, V. (2016). Blockchain technology: Beyond bitcoin. *Applied Innovation, 2*(6/10), 6–10.

Dadkhah, M., Rahimnia, F. and Filimonau, V. (2022). Evaluating the opportunities, challenges and risks of applying the blockchain technology in tourism: A Delphi study approach. *Journal of Hospitality and Tourism Technology*, *13*(5), 922–54.

Dogru, T., Mody, M. and Leonardi, C. (2018). Blockchain technology & its implications for the hospitality industry. *Boston University*, 1–12.

Elisa, N., Yang, L., Chao, F. and Cao, Y. (2018). A framework of blockchain-based secure and privacy-preserving e-government system. *Wireless Networks*, *29*(3), 1005-1015. https://doi.org/10.1007/s11276-018-1883-0.

Finck, M. (2018). Blockchain regulation and governance in Europe. Cambridge University Press.

Gangwal, A., Gangavalli, H.R. and Thirupathi, A. (2023). A survey of Layer-two blockchain protocols. *Journal of Network and Computer Applications*, *209*, 103539.

Golding, O., Yu, G., Lu, Q. and Xu, X. (2022, May). Carboncoin: Blockchain tokenization of carbon emissions with ESG-based reputation. In *2022 IEEE International Conference on Blockchain and Cryptocurrency (ICBC)* (pp. 1–5). IEEE.

Grover, P., Kar, A. and Janssen, M. (2019). Diffusion of blockchain technology. *Journal of Enterprise Information Management*, *32*(5), 735–57. https://doi.org/10.1108/jeim-06-2018-0132.

Guru, A., Mohanta, B.K., Mohapatra, H., Al-Turjman, F., Altrjman, C. and Yadav, A. (2023). A survey on consensus protocols and attacks on blockchain technology. *Applied Sciences*, *13*(4), 2604.

Hassani, H., Huang, X. and Silva, E. (2018). Banking with blockchain-ed big data. *Journal of Management Analytics*, *5*(4), 256–75. https://doi.org/10.1080/23270012.2018.1528900.

Hernandez, M.S., Sentosa, I., Gaudreault, F., Davison, I. and Sharin, F.H. (2023, May). The emergence of the metaverse in the digital blockchain economy: Applying the ESG framework for a sustainable future. In *2023 3rd International Conference on Advance Computing and Innovative Technologies in Engineering (ICACITE)* (pp. 1324–29). IEEE.

Jain, P., Singh, R.K., Mishra, R. and Rana, N. P. (2023). Emerging dimensions of blockchain application in tourism and hospitality sector: a systematic literature review. *Journal of Hospitality Marketing & Management*, *32*(4), 454–76.

Khanfar, A., Iranmanesh, M., Ghobakhloo, M., Senali, M. and Fathi, M. (2021). Applications of blockchain technology in sustainable manufacturing and supply chain management: a systematic review. *Sustainability, 13*(14), 7870. https://doi.org/10.3390/su13147870.

Kizildag, M., Dogru, T., Zhang, T.C., Mody, M.A., Altin, M., Ozturk, A.B. and Ozdemir, O. (2019). Blockchain: A paradigm shift in business practices. *International Journal of Contemporary Hospitality Management*, *32*(3), 953–75.

Kouhizadeh, M. and Sarkis, J. (2018). Blockchain practices, potentials, and perspectives in greening supply chains. Sustainability, 10(10), 3652.

Kumar, K.P.V., Lakshmi, B., Kumar, S.S., Muralidhar, V., Sai, N.R. and Nagamalleswara, V. (2023, August). Blockchain Technology: A Comprehensive Review and Future Directions.

In 2023 Second International Conference on Augmented Intelligence and Sustainable Systems (ICAISS) (pp. 1362–68). IEEE.

Kshetri, N. (2017). Can blockchain strengthen the internet of things?. IT professional, 19(4), 68–72.

Kwok, A.O. and Koh, S.G. (2019). Is blockchain technology a watershed for tourism development? *Current Issues in Tourism*, *22*(20), 2447–52.

Larcker, D.F., Pomorski, L., Tayan, B. and Watts, E.M. (2022). ESG ratings: A Compass without direction. *Rock Center for Corporate Governance at Stanford University Working Paper Forthcoming*.

Liu, F., Feng, Z. and Qi, J. (2022). A blockchain-based digital asset platform with multi-party certification. *Applied Sciences*, *12*(11), 5342.

Liu, X., Yang, Y., Jiang, Y., Fu, Y., Zhong, R.Y., Li, M. and Huang, G.Q. (2023). Data-driven ESG assessment for blockchain services: A comparative study in textiles and apparel industry. *Resources, Conservation and Recycling*, *190*, 106837.

Luong, N., Xiong, Z., Wang, P. and Niyato, D. (2018). Optimal auction for edge computing resource management in mobile blockchain networks: a deep learning approach. https://doi.org/10.1109/icc.2018.8422743.

Marr. B. (2018). Why blockchain could kill Uber. https://www.forbes.com/sites/bernardmarr/2018/02/09/why-blockchain-could-kill-uber/.

Mougayar, W. (2016). The business blockchain: Promise, practice, and application of the next internet technology. Wiley.

Modugu, K. (2020). Do corporate characteristics improve sustainability disclosure evidence from the UAE? *International Journal of Business Performance Management, 21*(1/2), 39. https://doi.org/10.1504/ijbpm.2020.106106.

Narayanan, A. (2016). Bitcoin and cryptocurrency technologies: a comprehensive introduction. Princeton University Press.

Nakamoto, S. and Bitcoin, A. (2008). A peer-to-peer electronic cash system. Bitcoin.–URL: https://bitcoin. org/bitcoin. pdf, 4(2), 15.

O'Dair, M. and Beaven, Z. (2017). The networked record industry: how blockchain technology could transform the record industry. *Strategic Change, 26*(5), 471–80. https://doi.org/10.1002/jsc.2147.

Ølnes, S., Ubacht, J. and Janssen, M. (2017). Blockchain in government: benefits and implications of distributed ledger technology for information sharing. *Government Information Quarterly, 34*(3), 355–364. https://doi.org/10.1016/j.giq.2017.09.007.

Ozdemir, A.I., Ar, I.M. and Erol, I. (2020). Assessment of blockchain applications in travel and tourism industry. *Quality & Quantity*, *54*, 1549–63.

Ozdemir, O., Dogru, T., Kizildag, M. and Erkmen, E. (2023). A critical reflection on digitalization for the hospitality and tourism industry: value implications for stakeholders. *International Journal of Contemporary Hospitality Management*.

Pal, A., Tiwari, C.K. and Haldar, N. (2021). Blockchain for business management: Applications, challenges and potentials. *The Journal of High Technology Management Research*, *32*(2), 100414.

Pazaitis, A., De Filippi, P. and Kostakis, V. (2017). Blockchain and value systems in the sharing economy: The illustrative case of Backfeed. Technological Forecasting and Social Change, 125, 105–115.

Peters, G.W. and Panayi, E. (2016). Understanding modern banking ledgers through blockchain technologies: Future of transaction processing and smart contracts on the internet of money. In P. Tasca, T. Aste, L. Pelizzon, & N. Perony (Eds.), Banking beyond banks and money (pp. 239–278). Springer.

Queiroz, M.M., Telles, R. and Bonilla, S.H. (2019). Blockchain and supply chain management integration: A systematic review of the literature. Supply Chain Management: An International Journal, 25(2), 241–254

Rabbani, M.R., Alshaikh, A., Jreisat, A., Bashar, A. and Moh'd Ali, M.A. (2021, December). Whether Cryptocurrency is a threat or a revolution? An analysis from ESG perspective. In *2021 International Conference on Sustainable Islamic Business and Finance* (pp. 103–108). IEEE.

Ronaghi, M. and Mosakhani, M. (2021). The effects of blockchain technology adoption on business ethics and social sustainability: Evidence from the middle east. *Environment Development and Sustainability, 24*(5), 6834–59. https://doi.org/10.1007/s10668-021-01729-x.

Saberi, S., Kouhizadeh, M., Sarkis, J. and Shen, L. (2019). Blockchain technology and its relationships to sustainable supply chain management. *International Journal of Production Research*, *57*(7), 2117–35.

Sandberg, H., Alnoor, A. and Tiberius, V. (2022). Environmental, social, and governance ratings and financial performance: evidence from the European food industry. *Business Strategy and the Environment, 32*(4), 2471–89. https://doi.org/10.1002/bse.3259.

Sharif, M. and Ghodoosi, F. (2022). The ethics of blockchain in organizations. *Journal of Business Ethics, 178*(4), 1009–1025. https://doi.org/10.1007/s10551-022-05058-5.

Swan, M. (2015). Blockchain: Blueprint for a new economy. "O'Reilly Media, Inc.".

Tapscott, D. and Tapscott, A. (2016). Blockchain revolution: how the technology behind bitcoin is changing money, business, and the world. Penguin.

Tapscott, D. and Tapscott, A. (2017). Realizing the potential of blockchain: A multistakeholder approach to the stewardship of blockchain and cryptocurrencies. World Economic Forum.

Treiblmaier, H. and Beck, R. (Eds.). (2019). *Business transformation through blockchain* (Vol. 1). Cham, Switzerland: Palgrave Macmillan.

Turgay, S. and Erdoğan, S. (2023). Enhancing Trust in Supply Chain Management with a Blockchain Approach. *Journal of Artificial Intelligence Practice*, *6*(6), 56–64.

Underwood, S. (2016). Blockchain beyond bitcoin. Communications of the ACM, 59(11), 15–17.

Upadhyay, A., Mukhuty, S., Kumar, V. and Kazancoglu, Y. (2021). Blockchain technology and the circular economy: implications for sustainability and social responsibility. *Journal of Cleaner Production, 293*, 126130.https://doi.org/10.1016/j.jclepro.2021.126130.

Wang, H., Zhang, M., Ying, H. and Zhao, X. (2021). The impact of blockchain technology on consumer behavior: a multimethod study. *Journal of Management Analytics, 8*(3), 371–90. https://doi.org/10.1080/23270012.2021.1958264.

World Economic Forum. (2017). *Digital Transformation Initiative: Aviation, Travel and Tourism Industry* [White paper]. Geneva.

Wu, H. and Zhu, X. (2020). Developing a reliable service system of charity donation during the COVID-19 outbreak. *IEEE Access, 8*, 154848–154860. https://doi.org/10.1109/access.2020.3017654.

Wu, W., Fu, Y., Wang, Z., Liu, X., Niu, Y., Li, B. and Huang, G.Q. (2022). Consortium blockchain-enabled smart ESG reporting platform with token-based incentives for corporate crowdsensing. *Computers & Industrial Engineering, 172*, 108456.

Wylde, V., Rawindaran, N., Lawrence, J., Balasubramanian, R., Prakash, E., Jayal, A., ... and Platts, J. (2022). Cybersecurity, data privacy and blockchain: a review. *SN Computer Science*, *3*(2), 127.

Yaqoob, I., Salah, K., Jayaraman, R. and Al-Hammadi, Y. (2022). Blockchain for healthcare data management: opportunities, challenges, and future recommendations. *Neural Computing and Applications*, 1–16.

Yucel, M. M., Baser, M.Y. and Kizildag, M. (2023). Distributed ledger technology in digital marketing: Enhancing customer relationships. Journal of Business Research, 148, 149–160.

Zhang, R., Xue, R. and Liu, L. (2019). Security and privacy on blockchain. *ACM Computing Surveys (CSUR)*, *52*(3), 1–34.

Zheng, Y. and Boh, W.F. (2021). Value drivers of blockchain technology: A case study of blockchain-enabled online community. *Telematics and Informatics*, *58*, 101563.

Zyskind, G. and Nathan, O. (2015, May). Decentralizing privacy: Using blockchain to protect personal data. In 2015 IEEE security and privacy workshops (pp. 180–184). IEEE.

3 Empowering Small-Scale Agriculture through Blockchain and Distributed Ledger Technology: Review and Future Perspectives

Sabrina Scuri[1*], Nuno Jardim Nunes[2] and Valentina Nisi[3]

1. Introduction

Agriculture plays an essential role in modern society and still represents the main source of livelihood for most people in low-income and developing countries. According to data provided by the World Bank[#] (World Bank, n.d.), the 2022 rate of employment in agriculture in low-income and least developed countries was respectively 58.8% and 54.2% (global rate was 26.3%). The food demand of a growing global population is now putting strains on small farmers worldwide, and especially those from low- and middle-income countries, who struggle to access markets (Tripoli and Schmidhuber, 2020). The current global food market is indeed dominated by large-scale producers and is therefore complex and risky for small players, as it requires the participation of several intermediaries (Kamilaris et al., 2019; Kumarathunga et al., 2022). Access to markets is not the only challenge farmers are facing. The agri-food supply chain is also characterised by transparency and efficiency issues (Tripoli and Schmidhuber, 2020). If on the one hand, consumers demand transparent information about their food purchases (Rocha et al., 2021), then on the other hand, the number of actors involved in the supply chain negatively impacts information flow, limiting traceability and quality control (Kamilaris et al., 2019).

In this context, Distributed Ledger Technologies (DLTs)—such as blockchains and smart contracts—are gaining increasing attention as they come with the promise of boosting the overall sector and promoting inclusive market participation (Natanelov et al., 2022). In particular, the application of DLTs is expected to bring

[1] Politecnico di Milano, Department of Design, Milan, Italy; ITI/LARSyS, Portugal.
[2,3] Instituto Superior Técnico-U. Lisbon, Departamento de Eng enharia Informática, Lisbon, Portugal; ITI/LARSyS, Portugal.
Email: sabrina.scuri@polimi.it[1]; nunojnunes@tecnico.ulisboa.pt[2]; valentina.nisi@tecnico.ulisboa.pt
* Corresponding author: sabrina.scuri@polimi.it

[#] Based on data sourced from the International Labour Organization modelled estimates database https://ilostat.ilo.org/data/

significant efficiency gains throughout the value chain by improving transparency and traceability (Tripoli and Schmidhuber, 2020) and redefining trust (Powell et al., 2023). Additional emerging technologies such as IoT, AI, and cloud computing—just to name some—are already being used to automatically collect food information along all stages of the supply chain, from production to distribution (Feng et al., 2020; Mondal et al., 2019; Powell et al., 2023; Tian, 2016, 2017). Traditional supply chain traceability systems are mostly centralised and "rely on a single information point to store, transmit, and share the traceability information" (Feng et al., 2020, p. 2), therefore, the data is vulnerable to tampering (Powell et al., 2022; Tian, 2017) and difficult for small farmers, consumers and other stakeholders to access (Xiong et al., 2020). As a distributed database system allowing for secure, transparent, and immutable recording of data, the application of DLTs is regarded as a solution to reinforce transparency, build trust across the supply chain (Feng et al., 2020; Tripoli and Schmidhuber, 2020), and ultimately improve traceability and quality control (Rocha et al., 2021; Tripoli and Schmidhuber, 2018).

Blockchain is therefore regarded as a promising solution to many challenges faced by the agri-food industry, particularly those affecting small farm holders. According to Kamilaris et al. (2019), blockchain adoption in the agricultural sector started gaining attention soon after the technology appeared, and the number of its applications is growing exponentially ever since (Rocha et al., 2021; Tribis et al., 2018). To date, companies and start-ups are running pilot projects all over the world to test the potential of such technology; yet, the success of these initiatives is still limited as most of them do not go beyond the Proof-of-Concept (PoC) stage (Kamilaris et al., 2019, Rocha et al., 2021; Tribis et al., 2018). Literature suggests several possible explanations for their limited success—these will be addressed in the section "Challenges and Possible Way Forward". Among them, we observe a research trend focused on technology development instead of real-world application (Scuri et al., 2023).

The hypothesis advanced in this chapter is that adopting a user-centred approach in the design of such solutions may help mitigate most of the existing barriers to success and represents a potential direction for development in the field. Here we call for shifting the current research focus from being tech-centred to being people-oriented and propose a set of design principles to inform future developments of DLT applications in the agrifood sector.

The reminder of our article is structured as follows. First, we provide an overview of prior works on DLT applications in the agri-food sector. We look at their main categories/purposes—(i) traceability, (ii) transaction and record keeping, and (iii) financing and access to credit—and explore how the key features of blockchain technology are being leveraged to address the shortcomings in the current agri-food industry. Second, we delve into the reasons for the limited success of DLT applications and elaborate on the potential contribution that a user-centred design approach may bring. Based on literature on user-centred design, co-design, and DLT, we then advance three key design principles for informing future research and practice. These principles are further illustrated through a contextual review of existing non-academic/commercial initiatives. The chapter concludes with a discussion of limitations and implications of the proposed approach.

2. Prior Works on the Application of DLTs in the Agri-food Sector

The integration of DLTs and other emerging technologies to improve traceability in the food supply chain has been widely tested. For instance, Borrero (2019) designed and pilot-tested a system for the traceability of the berry production chain in southern Spain. His Proof-of-Concept (PoC) uses smart contracts and permissioned ledgers to store food information (from field to manufacturing, packing, and shipping), ensuring that all actors share the same level of information. Mondal et al. (2019) propose an IoT and RFID blockchain-based architecture to track and provide consumers with a counterfeit-proof digital history of food packages' quality. Similarly, Tian (2016) created a traceability system based on blockchain and RFID to monitor food quality and safety 'from farm to fork.' The integration of DLTs in supply chain management has not only proved to be more reliable than traditional traceability systems, but also more efficient. In 2016, using IBM's blockchain solution, Walmart could trace a package of mangoes from the store to the source in less than three seconds—while it took the traditional system almost seven days to trace the origin of the fruits (Schilhabel et al., 2023).

Disintermediation of transactions and the use of smart contracts bring about significant financial benefits, particularly to small farmers and marginalised sections of rural communities in developing countries (Kamilaris et al., 2019; Tripoli and Schmidhuber, 2020). Smallholder farmers are often excluded from the market due to their low transaction volume (Xiong et al., 2020). Additionally, as family-size farms mostly rely on microtransactions, the use of traditional digital payment methods (like wire transfers or credit cards) is not always feasible because of the transaction fees (Kandeeban and Nivetha, 2019). As such, financial transactions in the agri-food sector are still heavily cash-based, therefore, they are not only slow and risky but also lack traceability, which ultimately hinders the ability to access credit (Tripoli and Schmidhuber, 2018). DLTs provide secure and real-time digital payment solutions that are more cost-effective than physical cash or even traditional digital payment methods (Tripoli and Schmidhuber, 2020; Xiong et al., 2020). Additionally, by moving to a DLT transaction system small farmers can create a tamper-proof history of transactions, which can then be used to prove their trustworthiness and access credit (Kumarathunga et al., 2022) or to inform production and marketing decisions (Espolov et al., 2020).

DLTs also provide a secure and efficient method for registering land titles to verify and secure private property, thereby addressing the shortcomings of traditional land registry systems, which are often inefficient, manual labour-based, and plagued by bureaucratic loopholes and opportunities for fraudulent behaviour (Alam et al., 2022). Assets recorded on the ledger—like immutable land titles—can be used by farmers as collateral, providing them with a further opportunity to access financing (Tripoli and Schmidhuber, 2020). Several projects deploying DLT systems to provide financial support to smallholder and third-world farmers have been launched so far[1]. For instance, in a study on the use of blockchain to promote financial resilience of Kenyan smallholders, Bolt (2019) identifies three main areas of application for the technology—(i) crop insurance, (ii) access to credit, and (iii) collaboration/pooling of resources—and illustrates them through nine use cases, most of them (six out of nine) belonging to the second category. One of such projects is *Agri-Wallet*, a mobile digital wallet for African farmers to sell products, save, and buy input supplies. As farmers receive funds and earnings in the form of tokens they can spend only within the agricultural supply chain, the blockchain-based solution works as an earmarked virtual currency thus creating a secure form of microfinancing. Furthermore, through *Agri-Wallet* farmers can get access to earmarked loans provided by the Rabobank Foundation (Bolt, 2019).

In their exploration on the use of blockchain in agriculture, Kim and Laskowski (2018) have also identified finance as one of the main application areas for this technology. In particular, the authors look at three existing applications—*AgriDigital*, *Etherisc*, and *Everex*—that leverage transparency and smart contract automation to benefit farmers in the developing world. Among them, *AgriDigital* is particularly worth mentioning as it has executed the world's first settlement of a physical commodity on a blockchain. *AgriDigital* is a blockchain-enabled platform that uses a set of smart contracts to support the trading, financing, and tracing of agri-commodities. Through the creation of digital assets, "AgriDigital brings together data flows that are often disparate in traditional, paper based agri-supply chains" (Sylvester, 2019, p. 25), and improves liquidity by providing users with a reliable prospect over their assets. Ultimately, by enabling real-time payments, *AgriDigital* removes the risk of buyers reneging on contracts, which happens often when market prices of the commodity purchased decrease making the contracts unprofitable. Kim and Laskowski (2018, p.13) argue that, "as payment is automatically scheduled and if the buyer actively stops payment, there is a well-documented evidence trail that can be used against them in litigation."

3. Challenges and Possible Way Forward

Despite the potential of DLT to radically transform the agri-food sector and provide the opportunity for more inclusive market participation for smallholder farmers, recent review works reveal an overall limited maturity of existing applications, suggesting that "convincing business cases are still scarce" (Kamilaris et al., 2019, p. 649). Rocha et al. (2021) argue that most of these initiatives are either PoC or laboratory prototypes. This claim is further supported by Tribis et al. (2018), who point out that around half of the solutions included in their systematic literature review are simply not ready for a real-world application, calling researchers to

1 An extensive list of such applications can be found in Bolt (2019), Kim & Laskowski (2018), and Sylvester (2019).

consider the feasibility and applicability of blockchain-based solutions in the agri-food sectors. A similar trend also emerges in the work by Kamilaris et al. (2019). Indeed, out of 49 projects analysed in their study, only four have reached the normal operations stage, while the majority (37) are still between the conceptual and the proof-of-concept stages or have become inactive.

According to the scholarly literature, there may be a number of explanations for the limited success rate of these initiatives. The most prominent (listed in Table 3.1) are briefly illustrated below. First, as most of the research on the subject is being conducted in the Computer Sciences and Engineering fields, the primary focus is still on technology development while real-world applications are less advanced (Rocha et al., 2021). Second, a key challenge lies in the limited adoption rate of DLT-based solutions by small- and medium- sized agricultural businesses. In fact, although these solutions are presented as a particularly important opportunity for smallholders and marginalised players, large-scale and fully operational projects are located in developed countries (Rocha et al., 2021) and run by big companies (Tripoli and Schmidhuber, 2018), which are likely to be supporting such initiatives because of "the hype of this technology" (Kamilaris et al., 2019, p. 70). As pointed out by Tripoli and Schmidhuber (2020), big market players are likely to obtain competitive advantage from exploiting DLTs; but smallholder farmers can gain significant benefits too, as long as the technology is made accessible to them. The digital divide is still an issue for rural agricultural communities all over the world—not only in developing countries (Marshall et al., 2020)—with issues like access to internet connections (Tripoli and Schmidhuber, 2020), technology availability, and interoperability of both technologies and data (Marshall et al., 2022), being the most prominent barriers to digital inclusion. Additionally, as the agricultural sector has never fully undergone a digital transformation (Tripoli and Schmidhuber, 2020), farmers often lack the time, resources and expertise to learn and take advantage of such cutting-edge technology (Kamilaris et al., 2019; Leduc et al., 2021).

Lack of understanding of how DLTs work, fuels scepticism and distrust (Schneider and Azan, 2022), thereby further exacerbating the adoption barrier and making it hard to convince farmers to give up on their old systems in favour of new DLT-based ones (Tribis et al., 2018). Previous research demonstrates that farmers' limited exposure and familiarity with advanced technologies—particularly in poor, rural and underserved communities (Drain et al., 2019)— is a challenge to designing and delivering effective solutions. Alongside the issues of technology literacy, a further emerging factor that seems to negatively affect the adoption of digital Agricultural Technologies (AgTech) solutions by farmers and other stakeholders, is the 'data divide' (Marshall et al., 2022). This concerns issues related to data governance, data literacy, privacy and trust (Marshall et al., 2022, Wiseman et al., 2019), which appear to be the main barrier to technology uptake.

Finally, we observe that the needs and challenges faced by small farmers are likely to be very context-specific, therefore, there is no one-size-fits-all solution. For instance, results from our recent study focused on Portuguese small farmers

(Scuri et al., 2023), suggest that financial benefits potentially brought by DLT applications might not be a strong motivation for their adoption by farmers living in a European middle-income country like Portugal. Furthermore, it should be noted that blockchain-based, and digital payment methods in general, were even regarded by our study participants as a limitation. On the contrary, by looking at the state-of-the-art, it emerges that most of the successful initiatives targeting farmers from the global south are centred on providing financial services and enabling mobile payments, which are indeed particularly popular in Asia and African regions like Kenya (Bolt, 2019; Leduc et al., 2021).

Table 3.1 Main reasons for the limited success of DLT applications in the agri-food sector

Most frequently mentioned reasons for the limited success of DLT applications in the agri-food sector	
1	Primary focus of existing initiatives is on technology development while real-world applications are less advanced.
2	Partial digital transformation of the sector and limited access to technology.
3	Complexity of DLT, limited technology and data literacy.
4	Solutions not sufficiently tailored to the specific local context and end-users needs

The many challenges to the success of DLT applications in the agricultural sector—in both developed and developing countries—are rooted in a plethora of social, technical, financial and regulatory hurdles. Among them, the literature highlights lack of access to robust telecommunications infrastructure (Marshall et al., 2020, 2022; Rotza et al., 2019; Salemink et al., 2017), limited provision of financial support and funding mechanisms (Tripoli and Schmidhuber, 2018; Sylvester, 2019), interoperability and compatibility of existing technologies (Marshall et al., 2022; Sylvester, 2019), fragmented and often inadequate regulatory frameworks for agricultural data management (Kamilaris, 2019; Tripoli and Schmidhuber, 2020; Wiseman et al., 2019), and ultimately the lack of education and training initiatives (Kamilaris, 2019; Marshall et al., 2022; Tripoli and Schmidhuber, 2018; Wiseman et al., 2019) as key challenges that farming communities all over the world are facing. Although acknowledging that these barriers need to be overcome in order to ensure success of DLT applications for the agri-food sector, here we advance that the lack of a user-centred approach in their design is the major hurdle to progress in the field. This corroborates earlier findings (Cao et al., 2022; Foth, 2017). Technology-enabled social innovation cannot be achieved only by focusing on technical and regulatory challenges. Due to the inherent complexity of DLT, understanding the social context in which the technology is going to be used is equally important to ensure that adoption is desirable and ethical. In a context that relies on limited access to technology, in order for blockchain to become an enabler for long-lasting positive impact, we need to provide much more than a technological solution. On the contrary, we should integrate the possibilities offered by the technology with the social context in which it is to be applied in order to provide an *enabling system*—i.e., a product or a service meant "to enable individuals and/or communities to achieve a result, using their skills and abilities while regenerating the quality of the living contexts in which they happen to live" (Manzini, 2007, p. 240).

User-Centred Design (UCD) and Design Thinking (DT) are "about people"; they're "about finding innovative solutions for people based on their needs" (Meinel and Leifer, 2015, p. 10). UCD and DT techniques have been long used to bolster innovation (Liedtka, 2015) by integrating people's needs with the possibilities of technology (Nash and Briggs, 2020). In this perspective, we argue that shifting from a tech-centred to a people-oriented approach would help address (or at least mitigate) most of the challenges described above.

Based on the analysis of existing research-based projects that adopt the suggested approach and recent Design and HCI literature on DLTs for agriculture, we put forward the following principles, hoping they may inspire and collectively serve as design guidelines for future research and practice:

- *Design Principle (DP) 1: Engaging users throughout the entire design process, from ideation to development.* Innovation in the agri-food sector cannot be achieved only by focusing on technology development. As suggested by Murray-Rust et al. (2023), alongside solving technical challenges, research on DLTs should seek to understand the social context in which the technology is going to be used and engage relevant stakeholders in devising the socio-technical possibilities being offered by these new systems. The value of engaging and dialoguing with end-users when developing blockchain-based solutions for the agri-food sector is also highlighted by Borrero (2019), Heitlinger et al. (2021), Pschetz et al. (2020), and Rudman (2020). Among the lessons learned from the deployment of Borrero's PoC, the author recommends researchers discuss possible use cases with relevant stakeholders so as to expand the agri-food blockchain ecosystem. Similarly, Pschetz and colleagues (2020) call for conducting contextual inquiry and interviews with multiple stakeholders to gather insights into their values, needs and situated conditions, and ultimately inform the design process. Heitlinger et al. (2021) take a step further. By using co-design and role play, the authors conducted a series of workshops with members of urban agricultural communities in London. In their project, the authors use blockchain as a 'design material' (Heitlinger et al., 2021) and adopt a post-human design perspective to envision alternative governance models and configurations of value within the food system for a fairer distribution of benefit, power and agency between human and non-human actors. In line with these research works, we argue that techniques such as immersive research, participatory design, design iterations with users, and testing are fundamental to ensure that the solutions being developed meet the specific needs of the target audience.
- *Design Principle (DP) 2: Facilitating connections between farmers and customers, and nurturing the sense of community.* Farmers often lack the

know-how and marketing skills to promote their businesses (Berti et al., 2017) and reach prospective customers (Dal Bello et al., 2022; Delgado, 2017). Although disintermediation and automation are presented as the main strengths of DLT, social and relational aspects are still of prominent importance. Community building has emerged as a potential driver for the adoption of DLTs applications, particularly in the energy sector (Scuri and Nunes, 2020; Scuri et al, 2019) and in the beef supply chain (McQueenie et al., 2024; Teli et al., 2022). A recent study (Scuri et al., 2023) extends these findings to family-size farming enterprises by showing that providing a solution for buyers to connect and get to know local producers favours engagement and creates a sense of community. Nurturing people-to-people relationships between consumers and producers was at the heart of BeefLegends, an award-winning design strategy aimed at connecting Australian beef producers with Chinese consumers (QUT, n.d.). By engaging both ends of the supply chain—i.e., producers and consumers—in the development of narrative content around food provenance and the local communities that produce it, BeefLegends augmented products data credentialling with meaningful local narratives and actively involved both producers and consumers "into new circuits of digital value creation" (McQueenie et al., 2021, p. 7). Thus, although DLTs are not necessarily needed for this purpose, we believe that enabling communication and exchange between users of such systems is an added value and could serve as a further driver towards adoption.

- *Design Principle (DP) 3: Demystifying blockchain, promoting technology understanding, and creating an enabling system through education and training*. Producing positive change by means of technology in a context like the one at hand, which relies on limited technology access, requires much more than simply providing an effective tool. To deliver on their promises, these initiatives should promote technology understanding, literacy training and provide support and onboarding services. The reputation surrounding blockchain technology, which has often been linked with online black-market activities (Sas and Khairuddin, 2017), and the general lack of understanding of how DLTs work, negatively affect adoption. Therefore, fighting misconceptions, promoting technology education, and ultimately creating an enabling environment for smallholder farmers to harness these new technologies (Tripoli and Schmidhuber, 2020) is fundamental for DLT applications in the agri-food sector to succeed.

4. Design Principles in Practice: A Contextual Review of Commercial Initiatives

To further illustrate the proposed design principles, we now present four initiatives as examples of their application. These initiatives were examined to complement the review of academic literature and research-based projects by extending the investigation to non-academic/commercial enterprise settings. We refer to this additional analysis as a 'contextual review', a term often used in design research to refer to the analysis of non-academic, contextual material (Gray and Malins, 2004), with the goal of grounding the investigation in the specific problem-context by integrating disciplinary/academic and experiential/non-academic knowledge (Barnes and Melles, 2007). The primary goal of this analysis is indeed to examine whether and how these principles have been applied to the development of commercial initiatives.

To identify these initiatives, we looked at published review articles (Abbas and Myeong, 2024; Kamilaris et al., 2019; Kim and Laskowski, 2018; Rocha et al., 2021; Tribis et al., 2018) as well as grey literature (Bolt, 2019; Sylvester, 2019), and checked references to identify additional cases. The initiatives analysed were selected as they met the following criteria:

- are built on some of the recommended design principles.
- are supported or run by companies that are still active[2].
- are commercial initiatives instead of pure research-based projects, as the latter tend to be small pilot studies or remain at a conceptual stage (Kamilaris et al., 2019).

As pointed out by other authors (Figueiredo and Reis, 2023; Kamilaris et al., 2019; Rudman, 2020), documentation about the status, outcomes and implementation of such initiatives is often incomplete or unavailable. We therefore had to rely on grey literature—when possible—and other non-academic sources—including the company's website, interviews to media outlets given by the company owners or other parties involved in the business, and any other document or dissemination material produced by the company. For this reason, the description provided of each initiative should not be considered comprehensive and, we acknowledge, may also be biased and not reflect the entire story. Nonetheless, we believe that the results of this complementary analysis are relevant as they do reflect the strategy and value proposition of the company.

4.1 AgUnity: A Human-centred Solution for 'Last-mile' Communities

Founded in 2016, *AgUnity* is a mobile platform powered by blockchain technology aimed at providing financial inclusion for communities of smallholder farmers worldwide. Initially pilot-tested through two studies conducted in Kenya and Papua New Guinea respectively (Sylvester, 2019), at the time of this writing the Australian start-up completed three projects and has seven more ongoing (AgUnity, n.d. a). As reported on the company's website, their approach is based on the assumption that "there are no assumptions." In fact, as stated by David Davies, CEO and one of the founders of *AgUnity*, "what works in Australia or other western countries tend not to translate in developing nations"; thus, gaining a complete

[2] The company operating status was determined based on Crunchbase data https://www.crunchbase.com/

understanding of users' needs was considered "critical to creating solutions that are relevant and truly work for farmers" (AgUnity, n.d. b). For this reason, the system has been developed following a user-centred design methodology that includes immersive research and participatory design (**DP 1**). To develop one of their first pilots, the team worked with members of remote farmer communities in Kenya and iteratively designed the platform together with them (AgUnity, n.d. b). As reported by Sylvester (2019), to ensure it could be easily understood and used by farmers with low levels of literacy or even illiterate, the final design of the *AgUnity* app was based on simplicity as a result of iterations. The company claims to believe in the importance of understanding specific challenges and opportunities of the people who are to use the platform, and has adopted diverse strategies to engage with the local communities, including partnering with NGOs and local farmers' cooperatives (Sylvester, 2019). *AgUnity* also addresses the lack of technology literacy as part of its mission (**DP 3**). The company aims to promote financial inclusion through digital inclusion, which it tries to achieve by: (i) simplicity of UX and UI—their platform "provides a suite of relevant tools that are easily comprehensible" (AgUnity, n.d. c); (ii) support—setting up of remote support teams as well as onboarding and guidelines materials (Asialink Business, n.d., York, 2021, Sylvester, 2019); and (iii) trust building. As part of its offering, the company provides for free both the infrastructure and basic smartphones pre-loaded with the app. Interestingly, although this business decision entails some costs for the company to bear, *AgUnity*'s business model remains commercially viable[3] and seems to be effective in overcoming initial technology adoption barriers (York, 2021). Besides supporting access to technology, trust-building is seen as another fundamental strategy to achieve actual adoption. As pointed out by David Davies "the farmer doesn't care about distributed ledger technology and all those things you're talking about, and they should not see the technology behind" (AgUnity, n.d. b). Thus, instead of leveraging on the technology, the creation of a direct relationship with farmers is presented as the way forward to demonstrate the company's willingness to tackle their needs and ultimately win their trust in the proposed solution.

4.2 1000EcoFarms: Connecting users and Building Communities

1000EcoFarms is a global B2B and B2C online marketplace for natural food, providing marketing, order, and payment services to farmers worldwide. The marketplace is based on FoodCoin, a global Ethereum-based blockchain ecosystem for food and agriculture products that uses smart contracts and its own cryptocurrency—called FoodCoin—to allow users to buy and sell goods on the *1000EcoFarms* platform (Motta et al., 2020). Although the use of this blockchain-based system seems to be the backbone of the platform, we could not find on the company website any reference to it or to blockchain technology in general. On the

[3] Further details can be found in International Finance Corporation & Mastercard Foundation (2018).

contrary, special emphasis is given to the mission of bringing "farmers and their products back to the center of the community" (1000EcoFarms, n.d.), a goal that the company tries to achieve by enabling social connections, communications, and information sharing between buyers and sellers (**DP 2**). *1000EcoFarms* seems to have shifted the focus from the enhanced capabilities offered by the technology to the value of being part of a community of like-minded people. In fact, as the CEO Mark Maytin claimed in an interview to the Mentor Capital Network (2017), "small farmers are not interested in technology—e.g. selling online—and will prefer to sell on farmers market". As "farmers don't want to deal with computer-related services," he proceeds, the company has changed its model by enabling the creation of 'food buying clubs' in which an administrator (usually a member of a neighbourhood community) acts as a small middleman responsible for managing joint orders for club members. Particularly interesting here is the fact that 'food buying clubs' are not simply presented to users as a convenient 'service,' but rather are described as communities of like-minded people interested in fresh local food and willing to share opinions and establish relationships with local producers. As information sharing and exchange is central to community building, *1000EcoFarms* also issues an online magazine that provides information about sustainable agriculture and includes a section 'Know your Farmer' that features stories of local food producers.

4.3 Arc-Net: Gaining Visibility and Business Insights

Another initiative meant to help farmers gain visibility and better promote their products is *Arc- Net*, a blockchain-based traceability solution for the food industry. Although the company's main value proposition is to bring transparency and security to food supply chains, end-user engagement seems to be among its main pillars (Arc-Net, n.d.). An integral part of Harvest ID—i.e., their solution for the food and farming industry—is indeed the Storybook. Storybooks are per-product interfaces farmers can use to showcase their businesses and share the history of their products (Samala, 2022) (**DP 2**). Storyboards can be customised and advanced on users' request. More advanced Storyboards, for instance, allow for richer consumer interactions and include additional features ranging from interactive content to gamification and two-way communication (Arc-Net, n.d.). This blockchain-based traceability system is described as highly technological since it integrates with other systems and technologies for data collection and access. Based on the description provided on the company's website, users are given the freedom to decide the amount of information to be shared as well as the data recording process (Arc-Net, n.d.). Moreover, *Arc-Net* gives producers access to data gathered from customers' interactions with Storyboards which can be analysed to identify purchasing patterns, get business intelligence data, and thus inform future marketing strategies (Samala, 2022). The number of services included in the system and the degree of customisation offered to end-users may represent a barrier for those who do not have much technical knowledge or lack marketing skills. As a possible workaround, *Arc-Net* offers support services throughout the journey, from onboarding to upgrade, to help guide producers in customising the solution to best fit their needs (**DP 3**).

4.4 Decapolis: Driving Positive Change through Training and Knowledge Sharing

Decapolis is a food safety and quality traceability platform based on blockchain. In this section we focus specifically on a pilot project aimed at empowering Jordanian smallholder farmers, which was launched in 2020 and developed in collaboration with the World Food Programme (WFP) Jordan Innovation Hub (Bandara, 2021; Decapolis, n.d.). In line with the mission of the WFP Jordan Innovation Hub (Nsour and Atieh, 2021), the proposed solution is context-specific, as it was designed to address challenges characterizing the Jordanian agri-food market and the needs of the local farming community (Decapolis, n.d.) (**DP 1**). In particular, the proposed solution aims to tackle two specific issues: (i) a lack of trust from the International Market towards quality standards of the Jordanian products (Devpost, n.d.), and (ii) the local farmers "lack of awareness of agricultural best practices and food safety standards" (Decapolis, n.d.). In the attempt to address the first issue, as claimed on the company's website and further stressed by the CEO (Devpost, n.d.), several actors and relevant stakeholders from both the public and private sector (e.g., government bodies, export regulators, industry organisations, food quality associations) were approached and engaged since the very beginning of project development to help validate their process. We couldn't find much information on the technical and regulatory aspects involved in the project. On the contrary, special emphasis is given to the solution proposed for addressing farmers lack of awareness of agricultural best practices and food safety standards. In this regard, together with the traceability platform, *Decapolis* provides farmers with training in the use of technology (Sudan and Rakhimova, 2022) as well as agricultural best practices (World Food Programme, n.d.) (**DP 3**).

Conclusion

The introduction of Distributed Ledger Technology in agriculture has the potential to radically transform the sector, enabling the creation of a more environmentally sustainable and socio-economically inclusive food system. Such a transformation is expected to particularly benefit small farm holders, bringing financial resilience, promoting trust and enabling access to global markets. Several companies and start-ups are running pilot projects all over the world to test the application of this promising technology in the agriculture and food industry. However, recent review works suggest that the success rate of the DLT-based projects launched so far is limited. Regulatory, financial and technical issues emerge as critical barriers towards adoption and the much-needed digital transformation of the food industry, enabled by blockchain technology. Governments, industry and the scientific community alike have recognized the need for overcoming these barriers and significant effort is already being made to address them. Nonetheless, in order for these initiatives to fully deliver on the promise of making positive impact in the sector, the social context in which the technology is going to be used needs to be carefully taken into account. Some challenges (as well as users' needs) are indeed very context-specific; therefore, a one-size-fits-all solution is quite unlikely to exist.

Our intent with this chapter is to fuel a critical discussion on the current approach to the design of DLT interventions for agriculture, suggesting the need for a paradigm shift. Here we argue that actual, long-term innovation within the sector and greater adoption of these systems can be achieved only by moving the focus from the technology to its users—that is, by adopting a user-centred approach. Through the analysis of recent Design and HCI literature on DLT and agriculture, combined with a contextual review of existing commercial initiatives, we put forward and illustrated three design principles to inform the adoption of the proposed approach within the industry and ultimately stimulate future research and practice.

First, we call for genuine codesign engagement with users and other relevant stakeholders before ideating and developing a new solution to get a deep understanding of the problem context and map out users' needs. As suggested by the case of *AgUnity*, getting such insights is fundamental to fostering adoption and trust. Moreover, this could help identify additional services and features to further boost adoption (for example, the training on agriculture best practices provided by *Decapolis*). Likewise, users should be involved throughout the entire design process. Multiple iterations with actual end-users should be conducted to ensure usability and ease-of-use of the system.

Second, we highlight the role of social connections in fostering adoption. Farmers often lack the marketing skills to promote their products and reach prospective customers. At the same time, customers are increasingly interested in buying local organic food. Given this perspective, a system that helps connect buyers and sellers—like *Arc-Net* and *1000EcoFarms*—would provide a relevant solution to a genuine problem. Moreover, by supporting social connections and interactions, it is possible to foster a sense of community among users and, as suggested by the case of *1000EcoFarms*, a system designed around community building is more likely to achieve actual and long-lasting engagement.

Third, in line with previous HCI research (Elsden et al., 2018; Scuri et al, 2019), we suggest addressing the technology's embedded complexity as a further potential barrier. On the one hand, effort should be devoted to conveying the technology in ways that can be easily understood by non-tech-savvy users, e.g., avoiding jargon, removing confusing terms (Bolt, 2019), or simplifying processes. On the other hand, to overcome this barrier it is fundamental to provide onboarding services (e.g., *AgUnity*), digital inclusion and data literacy training, and guidance to support the use of the system (e.g., *Arc-Net* and *Decapolis*).

The suggested principles are quite generic and have been elaborated based on the characteristics of markets and communities targeted by the works included in our analysis. Hence, they are not comprehensive; further principles may apply to other contexts. Likewise, the contextual analysis conducted to illustrate possible ways of applying such principles when devising blockchain-based initiatives is by no means exhaustive. The lack of detailed documentation on existing commercial initiatives represents a further limitation of our study. We couldn't find information about the adoption rate of the selected initiatives or clear evidence of market success. The only company that has released information regarding the impact made so far—which is based on measurable metrics such as (i) increase in revenue, (ii) compliance to safety and quality standards, (iii) waste reduction, and (iv) increase in productivity (Decapolis, n.d.)—is *Decapolis*. Nonetheless, the information provided by the company is not enough to determine its success. Therefore, we see these projects mainly as examples of possible positive directions rather than best practices. The research community—particularly from the field of Design and HCI—is experimenting with alternative, user-centred approaches to the development of DLT applications for the agri-food sector. With this chapter we want to second such a call and extend it to commercial settings. We see the initiatives reported here as indicative of an emerging interest from commercial enterprise settings in exploring alternative approaches to the development of blockchain-based solutions.

We believe that shifting from a tech-centred to a people-oriented approach may help in devising *enabling systems* (Manzini, 2007) rather than technological solutions; thereby bolstering innovation and producing positive impact on both small farming communities as well as on the industry. Nonetheless, this paradigm shift cannot be achieved without additional effort from the private and public sector, and the support of government-led training and capacity development initiatives. Adopting the suggested approach, indeed requires the development of additional skills—both technical/digital (Tripoli and Schmidhuber, 2020) and "'softer-skills' in community engagement and digital marketing" (McQueenie et al., 2021)—that SMEs interested in developing these solutions will need to acquire. Another possible way forward to help commercial initiatives to take off is strengthening the already-existing collaboration and mutual exchange between industry and academia. The combination of academic and non-academic knowledge production—which we refer to as transdisciplinary research (Choi and Pak, 2006)—helps address challenges in a holistic way and, therefore, is particularly relevant to the development of DLT applications in agri-food sector.

Acknowledgements

This research work has received funding from the European Union's Horizon 2020 research and innovation programme under grant agreement no 952226.

References

1000EcoFarms. (n.d.). *About Us*. https://www.1000ecofarms.com/en/promo/seller.

Abbas, Z. and Myeong, S. (2024). A Comprehensive Study of Blockchain Technology and Its Role in Promoting Sustainability and Circularity across Large-Scale Industry. *Sustainability*, *16*(10), 4232. https://doi.org/10.3390/su16104232.

AgUnity. (n.d. a). *Our Projects*. https://www.agunity.com/projects.

AgUnity. (n.d. b). *How AgUnity is Changing the Developing World with Technology in the Hands of Farmers*. https://www.agunity.com/blog/how-agunity-is-changing-the-developing- world-with-technology-in-the-hands-of-farmers.

AgUnity. (n.d. c). *Our Digital Solutions*. https://www.agunity.com/solutions.

Alam, K.M., Rahman, J.A., Tasnim, A. and Akther, A. (2022). A blockchain-based land title management system for Bangladesh. *Journal of King Saud University-Computer and Information Sciences, 34*(6), 3096-3110. https://doi.org/10.1016/j.jksuci.2020.10.011.

Arc-Net. (n.d.). *arc-net. Food & Farming*. https://www.arc-net.io/.

Asialink Business. (n.d.). *AgUnity Case Study: Accelerating AgTech Australia's opportunity in Vietnam*. https://asialinkbusiness.com.au/news-media/agunity-case-study.

Bandara, K.Y. (2021). *A Detailed Analysis of Blockchain Pilots and Applications in Highly Impacted Areas*. https://assets.gov.ie/280153/0496539d-698d-4968-8ea8-94dbece28030.pdf.

Barnes, C. and Melles, G. (2007). Managing interdisciplinarity: a discussion of the contextual review in design research. *Proceedings of the International Association of Societies of Design Research Conference (IASDR) 2007*. Hong Kong Polytechnic University.

Berti, G., Mulligan, C. and Yap, H. (2017). Digital food hubs as disruptive business models based on Coopetition and "shared value" for sustainability in the agri-food sector. In S. Sindakis & P. Theodorou (Eds.), *Global Opportunities for Entrepreneurial Growth: Coopetition and Knowledge Dynamics within and across Firms* (pp. 415–38). Emerald Publishing Limited. https://doi.org/10.1108/978-1-78714-501-620171023.

Bolt, J.S. (2019). *Financial resilience of Kenyan smallholders affected by climate change, and the potential for blockchain technology*. CCAFS. http://library.wur.nl/WebQuery/wurpubs/fulltext/472583.

Borrero, J.D. (2019). Agri-food supply chain traceability for fruit and vegetable cooperatives using blockchain technology. *Revista de Economía Pública, Social y Cooperativa*, *95*, 71–94. https://doi.org/10.7203/CIRIEC-E.95.13123.

Cao, S., Foth, M., Powell, W., Miller, T. and Li, M. (2022). A blockchain-based multisignature approach for supply chain governance: A use case from the Australian beef industry. *Blockchain: Research and Applications*, *3*(4), 100091. https://doi.org/10.1016/j.bcra.2022.100091.

Chandrakumar, C., McLaren, S.J., Jayamaha, N.P. and Ramilan, T. (2019). Absolute sustainability-based life cycle assessment (ASLCA): a benchmarking approach to operate agri-food systems within the 2 C global carbon budget. *Journal of Industrial Ecology, 23*(4), 906–917. https://doi.org/10.1111/jiec.12830.

Choi, B.C. and Pak, A.W. (2006). Multidisciplinarity, interdisciplinarity and transdisciplinarity in health research, services, education and policy: 1. Definitions, objectives, and evidence of effectiveness. *Clinical & Investigative Medicine*, *29*(6), 351–64.

Dal Bello, U.B., Marques, C., Sacramento, O. and Galvão, A. (2022). Neo-rural small entrepreneurs' motivations and challenges in Portugal's low density regions. *Journal of Enterprising Communities: People and Places in the Global Economy*, *16*(6), 900–23. https://doi.org/10.1108/JEC-04-2021-0047.

Decapolis. (n.d.). *WFP Case Study. Blockchain Traceability Platform for Smallholder Farmers*. https://www.decapolis.ai/wfp-case-study.

Delgado, C. (2017). Mapping urban agriculture in Portugal: Lessons from practice and their relevance for European post-crisis contexts. *Moravian Geographical Reports*, *25*(3), 139–53. https://doi.org/10.1515/mgr-2017-0013.

Devpost. (n.d.). *Decapolis provides a platform for certification of food quality to food producers to providing safe goods for which the end-to-end supply chain complies with strict quality control standards*. https://devpost.com/software/decapolis.

Drain, A., Shekar, A. and Grigg, N. (2019). "Involve me and I'll understand": creative capacity building for participatory design with rural Cambodian farmers. *CoDesign*, *15*(2), 110–127. https://doi.org/10.1080/15710882.2017.1399147.

Elsden, C., Manohar, A., Briggs, J., Harding, M., Speed, C. and Vines, J. (2018). Making sense of blockchain applications: A typology for HCI. In *Proceedings of the 2018 chi conference on human factors in computing systems* (pp. 1–14). https://doi.org/10.1145/3173574.3174032.

Espolov, T., Espolov, A., Kalykova, B., Umbetaliyev, N., Uspanova, M. and Suleimenov, Z. (2020). Asia agricultural market: Methodology for complete use of economic resources through supply chain optimization. *International Journal of Supply Chain Management, 9*(3), 408. https://ojs.excelingtech.co.uk/index.php/IJSCM/article/view/4912.

Feng, H., Wang, X., Duan, Y., Zhang, J. and Zhang, X. (2020). Applying blockchain technology to improve agri-food traceability: A review of development methods, benefits and challenges. *Journal of cleaner production, 260*, 121031. https://doi.org/10.1016/j.jclepro.2020.121031.

Figueiredo, B. and Reis, C.I. (2023). COW: A Proof-of-Concept DLT Platform for the Agri-Food Supply Chain. In *2023 IEEE International Conference on Artificial Intelligence, Blockchain, and Internet of Things (AIBThings)* (pp. 1-5). IEEE. https://doi.org/10.1109/AIBThings58340.2023.10291025.

Foth, M. (2017). The promise of blockchain technology for interaction design. *Proceedings of the 29th Australian Conference on Computer-Human Interaction* (pp. 513–517). https://doi.org/10.1145/3152771.3156168.

Gray, C. and Malins, J. (2004). *Visualizing Research: A Guide to the Research Process in Art and Design*. Ashgate.

Heitlinger, S., Houston, L., Taylor, A. and Catlow, R. (2021). Algorithmic Food Justice: Co-Designing More-than-Human Blockchain Futures for the Food Commons. In *CHI '21: Proceedings of the 2021 CHI Conference on Human Factors in Computing Systems* (pp. 1-17). https://doi.org/10.1145/3411764.3445655.

ILO. (n.d.). *Indicators and data tools*. https://ilostat.ilo.org/data/.

International Finance Corporation, & Mastercard Foundation. (2018). *Handbook: Digital Financial Services for Agriculture*. https://www.ifc.org/content/dam/ifc/doc/mgrt/digital- financial-services-for-agriculture-ifc-mcf-2018.pdf.

Kamilaris, A., Fonts, A. and Prenafeta-Boldú, F.X. (2019). The rise of blockchain technology in agriculture and food supply chains. *Trends in food science & technology, 91*, 640–52. https://doi.org/10.1016/j.tifs.2019.07.034.

Kandeeban, M. and Nivetha, T. (2019). Blockchain: A tool for a secure, safe and transparent way of food and agricultural supply chain. *International Journal of Farm Sciences, 9*(1), 97–100. https://doi.org/10.5958/2250-0499.2019.00025.9.

Kim, H.M. and Laskowski, M. (2018). *Agriculture on the blockchain: Sustainable solutions for food, farmers, and financing*. Supply Chain Revolution, Barrow Books. https://papers.ssrn.com/sol3/Delivery.cfm/SSRN_ID3028164_code2969338.pdf?abstractid=3 028164&mirid=1.

Kumarathunga, M., Calheiros, R.N. and Ginige, A. (2022). Smart agricultural futures market: Blockchain technology as a trust enabler between smallholder farmers and buyers. *Sustainability, 14*(5), 2916. https://doi.org/10.3390/su14052916.

Leduc, G., Kubler, S. and Georges, J.P. (2021). Innovative blockchain-based farming marketplace and smart contract performance evaluation. *Journal of Cleaner Production, 306*, 127055. https://doi.org/10.1016/j.jclepro.2021.127055.

Liedtka, J. (2015). Perspective: Linking design thinking with innovation outcomes through cognitive bias reduction. *Journal of product innovation management*, *32*(6), 925–38. https://doi.org/10.1111/jpim.12163.

Manzini, E. (2007). Design research for sustainable social innovation. In R. Michel (Ed.), *Design research now: Essays and selected projects* (pp. 233–45). Birkhäuser Basel. https://doi.org/10.1007/978-3-7643-8472-2_14.

Marshall, A., Turner, K., Richards, C., Foth, M. and Dezuanni, M. (2022). Critical factors of digital AgTech adoption on Australian farms: from digital to data divide. *Information, Communication and Society*, *25*(6), 868–86. https://doi.org/10.1080/1369118X.2022.2056712.

Marshall, A., Dezuanni, M., Burgess, J., Thomas, J. and Wilson, C.K. (2020). Australian farmers left behind in the digital economy–Insights from the Australian Digital Inclusion Index. *Journal of Rural Studies*, *80*, 195–210. https://doi.org/10.1016/j.jrurstud.2020.09.001.

McQueenie, J., Foth, M. and Hearn, G. (2024). *Community, Culture, Commerce: The Intermediary in Design and Creative Industries*. Palgrave Macmillan. https://doi.org/10.1007/978-981-99-7889-2.

McQueenie, J.M., Foth, M., Powell, W. and Hearn, G. (2021). *BeefLegends: Connecting the Dots between Community, Culture and Commerce*. QUT Design Lab, Brisbane. https://doi.org/10.5204/rep.eprints.213769.

Meinel, C. and Leifer, L. (2015). Introduction–Design thinking is mainly about building innovators. Design thinking research: Building innovators. In H. Plattner, C. Meinel, & L. Leifer, L. (Eds.), *Design Thinking Research. Understanding Innovation* (pp. 1–11). Springer, Cham. https://doi.org/10.1007/978-3-319-06823-7_1.

Mentor Capital Network. (2017, April 4). *Interview: 1000 EcoFarms*. https://mentorcapitalnet.org/2017/04/04/interview-1000-ecofarms/.

Mondal, S., Wijewardena, K.P., Karuppuswami, S., Kriti, N., Kumar, D. and Chahal, P. (2019). Blockchain inspired RFID-based information architecture for food supply chain. *IEEE Internet of Things Journal, 6*(3), 5803–5813. https://doi.org/10.1109/JIOT.2019.2907658.

Motta, G.A., Tekinerdogan, B. and Athanasiadis, I.N. (2020). Blockchain applications in the agri-food domain: the first wave. *Frontiers in Blockchain, 3*, 6. https://doi.org/10.3389/fbloc.2020.00006.

Murray-Rust, D., Elsden, C., Nissen, B., Tallyn, E., Pschetz, L. and Speed, C. (2023). Blockchain and beyond: Understanding blockchains through prototypes and public engagement. *ACM Transactions on Computer-Human Interaction, 29*(5), 1–73. https://doi.org/10.1145/3503462.

Nash, C. and Briggs, J. (2020). All Innovation is Social. In *Design revolutions: IASDR 2019 Conference Proceedings. Volume 2: Living, Making, Value* (pp. 547–65). https://nrl.northumbria.ac.uk/id/eprint/39828/.

Natanelov, V., Cao, S., Foth, M. and Dulleck, U. (2022). Blockchain smart contracts for supply chain finance: Mapping the innovation potential in Australia-China beef supply chains. *Journal of Industrial Information Integration*, *30*, 100389. https://doi.org/10.1016/j.jii.2022.100389.

Nsour, A. and Atieh, F. (2021, January 19). Innovative Jordan: WFP's innovation hub is accelerating Jordanian startups. *Medium*. https://wfpinnovation.medium.com/innovative-jordan-wfps-innovation-hub-is-accelerating-jordanian-startups-66de74f0380c.

Powell, W., Cao, S., Foth, M., He, S., Turner-Morris, C. and Li, M. (2023). Revisiting Trust in Supply Chains: How Does Blockchain Redefine Trust? In A. Bouras, I. Khalil, & B. Aouni (Eds.), *Blockchain Driven Supply Chains and Enterprise Information Systems* (pp. 21–42). Springer International Publishing. https://doi.org/10.1007/978-3-030-96154-1_2.

Powell, W., Foth, M., Cao, S. and Natanelov, V. (2022). Garbage in garbage out: The precarious link between IoT and blockchain in food supply chains. *Journal of Industrial Information Integration*, *25*, 100261. https://doi.org/10.1016/j.jii.2021.100261.

Pschetz, L., Dixon, B., Pothong, K., Bailey, A., Glean, A., Soares, L.L. and Enright, J.A. (2020, April). Designing distributed ledger technologies for social change: the case of CariCrop. In *Proceedings of the 2020 CHI conference on human factors in computing systems* (pp. 1–12). ACM. https://doi.org/10.1145/3313831.3376364.

QUT. (n.d.). *BeefLedger: 2020 Good Design Awards Winner*. https://research.qut.edu.au/best/beefledger-2020-good-design-awards-winner/.

Rocha, G.D.S.R., de Oliveira, L. and Talamini, E. (2021). Blockchain applications in agribusiness: A systematic review. *Future Internet, 13*(4). https://doi.org/10.3390/fi13040095.

Rudman, H. (2020). Piece of Cake: Assuring Specific Qualities of Product in Farm Lifecycles with DLT – Can Evidenced Based Practice be supported by Participatory Action Research Methods? *The Journal of The British Blockchain Association*, *4*(1), 1–7. https://doi.org/10.31585/jbba-4-1-(1)2021.

Samala, H. (2022). Blockchain technology in Agriculture: Applications, Impact and Future. *International Research Journal of Engineering and Technology*, *9*(11), 77–83.

Sas, C. and Khairuddin, I.E. (2017). Design for trust: An exploration of the challenges and opportunities of bitcoin users. In *Proceedings of the 2017 CHI Conference on Human Factors in Computing Systems* (pp. 6499-6510). ACM. https://doi.org/10.1145/3025453.3025886.

Schilhabel, S., Sankaranarayanan, B., Basu, C., Madan, M., Glennan, C. and McSherry, L. (2023). Blockchain technology in the food supply chain: influences on supplier relationships and outcomes. *Issues in Information Systems, 24*(3), 321–32.

Schneider, B. and Azan, W. (2022, June). Perceptions and Misconceptions of Blockchain: The Potential of Applying Threshold Concept Theory. In *2022 IEEE 6th International Conference on Logistics Operations Management (GOL)*, (pp. 1-6). IEEE. https://doi.org/10.1109/GOL53975.2022.9820452.

Scuri, S., Ribeiro, C. and Nisi, V.(2023) A design-driven approach to distributed ledger technologies for small farmers communities: a case study in Portugal. In D. De Sainz Molestina, L. Galluzzo, F. Rizzo, & D. Spallazzo (Eds.), *IASDR 2023: Life-Changing Design*. Design Research Society. https://doi.org/10.21606/iasdr.2023.217.

Scuri, S. and Nunes, N.J. (2020). PowerShare 2.0: A Gamified P2P Energy Trading Platform. In *Proceedings of the International Conference on Advanced Visual Interfaces* (pp. 1–3). ACM. https://doi.org/10.1145/3399715.3399948.

Scuri, S., Tasheva, G., Barros, L. and Nunes, N.J. (2019). An hci perspective on distributed ledger technologies for peer-to-peer energy trading. In D. Lamas, F. Loizides, L. Nacke, H. Petrie, M. Winckler, & P. Zaphiris (Eds.), *Human-Computer Interaction – INTERACT 2019. INTERACT 2019*. Lecture Notes in Computer Science, vol 11748 (pp. 91–111). Cham: Springer International Publishing. https://doi.org/10.1007/978-3-030-29387-1_6.

Singh, V. and Sharma, S.K. (2022). Application of blockchain technology in shaping the future of food industry based on transparency and consumer trust. *Journal of Food Science and Technology*, 1–18. https://doi.org/10.1007/s13197-022-05360-0.

Sudan, S. and Rakhimova, G. (2022, August 16). How regional innovation can transform the future of food systems. *Medium*. https://wfpinnovation.medium.com/how-regional-innovation-can-transform-the-future-of-food-systems-9013aaf84b33.

Sylvester, G. (2019). *E-agriculture in action: blockchain for agriculture, opportunities and challenges*. FAO. https://www.fao.org/documents/card/en?details=CA2906EN.

Teli, M., McQueenie, J., Cibin, R. and Foth, M. (2022). Intermediation in design as a practice of institutioning and commoning. *Design Studies*, *82*, 101132. https://doi.org/10.1016/j.destud.2022.101132.

Tian, F. (2017). A supply chain traceability system for food safety based on HACCP, blockchain & Internet of things. In *2017 International conference on service systems and service management* (pp. 1–6). IEEE. https://doi.org/10.1109/ICSSSM.2017.7996119.

Tian, F. (2016, June). An agri-food supply chain traceability system for China based on RFID & blockchain technology. In *2016 13th international conference on service systems and service management (ICSSSM)*, (pp. 1–6). IEEE. https://doi.org/10.1109/ICSSSM.2016.7538424.

Tribis, Y., El Bouchti, A. and Bouayad, H. (2018). Supply chain management based on blockchain: A systematic mapping study. In *MATEC Web of Conferences 200*(00020). https://doi.org/10.1051/matecconf/201820000020.

Tripoli, M. and Schmidhuber, J. (2020). *Emerging opportunities for the application of blockchain in the agri-food industry. Revised version.* FAO and ICTSD. https://www.fao.org/3/ca9934en/CA9934EN.pdf.

Tripoli, M. and Schmidhuber, J. (2018). *Emerging Opportunities for the Application of Blockchain in the Agri-food Industry*. FAO and ICTSD. https://ageconsearch.umn.edu/record/320187/files/Tripoli%20Schmidhuber%20blockchain% 202018.pdf.

Wiseman, L., Sanderson, J., Zhang, A. and Jakku, E. (2019). Farmers and their data: An examination of farmers' reluctance to share their data through the lens of the laws impacting smart farming. *NJAS-Wageningen Journal of Life Sciences*, *90-91*, 100301. https://doi.org/10.1016/j.njas.2019.04.007.

World Bank. (n.d.). *Employment in agriculture (% of total employment) (modeled ILO estimate)*. https://data.worldbank.org/indicator/SL.AGR.EMPL.ZS?contextual=default&end=2022&name_desc=false&skipRedirection=true&start=2022&type=shaded&view=map&year=2022.

World Food Programme. (n.d.). *Decapolis. Blockchain traceability platform for smallholder farmers.* https://innovation.wfp.org/project/decapolis.

Xiong, H., Dalhaus, T., Wang, P. and Huang, J. (2020). Blockchain technology for agriculture: applications and rationale. *Frontiers in Blockchain, 3*(7). https://doi.org/10.3389/fbloc.2020.00007.

York, W. (2021, October 27). *Australian start-up reaching last-mile communities*. AusBiz. https://ausbizmedia.com/agunity/.

4 Enhancing Healthcare System Performance Assessment through Blockchain Integration

Rami Alkhudary[1*], Małgorzata Macuda[2], Stessy Coutelle[3] and Pierre Féniès[4]

1. Introduction

The concept of health is multifaceted and holds an important position due to its magnitude in relation to human life. The establishment of a resilient healthcare system necessitates careful consideration of multiple aspects, including but not limited to life expectancy, productivity, healthcare requirements, and the satisfaction of numerous individual and community responsibilities (Allen-Duck et al., 2017). The healthcare industry possesses distinct attributes that distinguish it from other sectors. The principal aim of the healthcare ecosystem is to provide standard medical services to each member of the population while also ensuring the prudent deployment of public expenditures. Healthcare systems comprise a wide range of activities, such as pain treatment, disability mitigation, restoration of functionality, and disease prevention (Fuchs, 2018).

The healthcare industry operates in a regulatory environment characterized by constant change and evolution. It faces financial constraints and increasing competition, all while being continuously influenced by technological advances (Beaulieu et al., 2023; Cimasi, 2004). Healthcare supply chains confront challenges such as resource scarcity, a dynamic regulatory landscape, and various social, economic, political, and geographic factors. These value chains must not only deliver excellence but also provide cost-effective services. The concept of effectiveness includes several components, such as accessibility, continuity, and technical excellence (Alonazi and Thomas, 2014; Kludacz-Alessandri et al., 2021; Shortell, 1989). Furthermore, demographic shifts have resulted in a higher proportion of elderly individuals, notably in North America and the European Union. These demographic changes impose increased financial obligations on the public sector and necessitate that governments address the challenges posed by the rapid expansion of the healthcare sector within their social systems (Aydin, 2022).

[1,3,4] Université Paris-Panthéon-Assas, LARGEPA, Paris, France.
[2] Uniwersytet Ekonomiczny w Poznaniu
Email: malgorzata.macuda@ue.poznan.pl[2]; stessy.coutelle@u-paris2.fr[3]; pierre.fenies@u-paris2.fr[4]
* Corresponding author: rami.alkhudary@u-paris2.fr

The healthcare sector faces the complex task of identifying and valuing its performance. Assessing the demand for healthcare services is equally challenging. The escalation of healthcare costs can be attributed to the impact of rapid technological advancements and ongoing research and development in medical science (Fuchs, 2018). The core purpose of economic calculations should be to mitigate waste and mismanagement while optimizing the use of taxpayers' funds to equitably distribute limited resources in response to the demand for medical services. The concept of productivity encompasses the interplay between various inputs, such as resources, which encompass labor, capital, and equipment; outputs, including the number of patients treated and waiting times; and health outcomes (Aydin, 2022). Healthcare management will be compelled to reallocate resources due to the pressures arising from discussions on cost and performance (Benzidia et al., 2019).

Healthcare information systems possess limited capabilities to address the challenges associated with data exchange (Rejeb et al., 2023). Despite various restrictions and access controls, the security of medical data remains compromised, leading to a widespread lack of trust among a range of stakeholders. This mistrust hampers the benefits that could result from identifying and implementing best business practices to enhance health system performance. Moreover, the absence of data integration mechanisms acts as a barrier to achieving healthcare interoperability. Blockchain technology has evolved into a unique information system characterized by a decentralized and immutable database (Alkhudary and Féniès, 2022; Lacity and Treiblmaier, 2022). Academic literature underscores the potential of this technology for decentralized healthcare applications (Al-Sumaidaee, Alkhudary, Zilic, et al., 2023). These applications encompass monitoring and automating health claims processing, collecting, and sharing patient medical data, combating drug counterfeiting, and facilitating medical clinical trials. In order to improve the health system performance, the development of blockchain applications aimed at optimizing hospital performance should be prioritized (Al-Sumaidaee et al., 2021).

This chapter provides insights into how blockchain can be integrated into the Health System Performance Assessment (HSPA) framework, addressing key deficiencies in healthcare, data security, and resource management. Blockchain is notable for its potential to increase process transparency, enhance the security of sensitive patient data, and improve operational performance. The chapter explores how this technology can lead to more reliable and efficient healthcare services. It emphasizes that blockchain's capabilities extend beyond mere technical improvements, ensuring a robust framework for managing patient data and resource allocation. Furthermore, the chapter suggests that the broader implications of blockchain utilization in healthcare could lead to societal benefits. Enhanced trust in healthcare processes, driven by blockchain's immutable and transparent nature, encourages a more engaged and informed patient population. Additionally, the move toward more patient-centered services, supported by blockchain's ability to manage and share information, aligns healthcare outcomes with individual needs. The chapter highlights blockchain's role not merely as a technological upgrade but as a transformative force for healthcare systems and societal health governance.

As for the organization of the chapter, it begins with an Introduction. The Health System Performance Assessment (HSPA) Framework section follows, detailing the elements used to evaluate healthcare systems. The next section, Blockchain in Healthcare Enhancing Business Practices, explores the technical aspects of blockchain and how it can improve business practices in healthcare. This is followed by Enhancing HSPA Framework Implementation with Blockchain, which delves into how blockchain can improve the HSPA framework by providing transparent, secure, and accurate data management. The chapter then addresses areas for Future Research, highlighting directions for further exploration. Finally, the Conclusion and Limitations section summarizes the findings, discusses the implications, and the limitations.

2. Health System Performance Assessment (HSPA) Framework

Performance management is a critical prerequisite for the successful implementation of the management process in hospitals. It plays a pivotal role in ensuring the uninterrupted operations of hospitals and in cultivating a competitive edge. In the realm of healthcare, this is a complex undertaking, as it necessitates the evaluation of a multitude of goals set by hospitals, encompassing medical, social, and economic dimensions (Noto et al., 2019). Diligent consideration and the selection of an appropriate metric to assess performance are of paramount importance (KirungaTashobya et al., 2018).

Hence, it is imperative to develop an information system for the measurement and evaluation of performance. This system should be equipped to collect, record, and analyze data related to a hospital's outcomes. It should also encompass the aggregation and comparison of this data with predetermined objectives and standards and facilitate the communication of results (Bernal-Delgado et al., 2015). The system should be designed to accommodate the various variables inherent to each healthcare facility, while also addressing the informational needs of its management and other stakeholders (Barbazza et al., 2019). The Health System Performance Assessment (HSPA) framework for Universal Health Coverage (UHC) is presented in Figure 4.1. The framework facilitates the establishment of a broader network of connections between functions and their resulting consequences.

The concept of governance within the health system encompasses multiple sub-functions aimed at fostering collaboration across diverse sectors and employing legislation and regulations to advance public health objectives (Ntr et al., 2019; Smith et al., 2012). Its core objective is to exert an influence on the determinants of health, typically through collaborative efforts with other sectors. Health financing plays a pivotal role, involving the allocation and utilization of financial resources for healthcare delivery, thereby ensuring universal access to both public health services and individual healthcare. This includes decisions related to payment structures, provider compensation, service costs, and the range of services accessible to individuals. The process of resource generation guarantees that the healthcare system maintains access to all the necessary inputs for its operation. These inputs encompass a wide array of elements, including medical professionals, medical equipment, infrastructure, medications, immunizations, and various medical supplies. The primary function of resource generation lies in managing the production, procurement, availability, and maintenance of these inputs to meet healthcare service demands in a timely and appropriate manner. Additionally, service delivery, another facet of health systems, is intertwined with governance, financing, and resource generation. This concept involves the transformation of these inputs within a specific organizational framework, resulting in the delivery of a diverse range of healthcare solutions. The delivery of services impacts the intermediate goals of the health system and the achievement of the overarching health system objectives. This process entails the integration of various resources to deliver interventions within a well-defined organizational structure.

Fig. 4.1 HSPA Framework for UHC, adopted from Papanicolas et al. (2022)

The concept of effectiveness in healthcare refers to the extent to which a service successfully achieves its intended outcomes (Sanford et al., 2024), whether at the level of patients, the population, or within the healthcare system (Hargett et al., 2017). Safety is focused on the extent to which healthcare processes mitigate undesirable outcomes that may result from underlying healthcare processes. The domain of user experience involves the assessment and evaluation of the perspective and overall experience of recipients of healthcare services, emphasizing their significance as a critical outcome of service delivery. Lastly, access refers to the extent to which health services are not only readily available but also easily accessible within a reasonable time frame while ensuring adequate financial protection for individuals. Collectively, these objectives establish a comprehensive framework for evaluating the quality of healthcare ecosystems.

The concept of financial protection in healthcare pertains to the healthcare system's capacity to shield individuals from the financial risks associated with illness or healthcare expenditures (Singh and Barakoti, 2023). This is achieved through the implementation of health insurance mechanisms and the establishment of comprehensive universal health coverage, ensuring that individuals are not financially devastated by health-related costs (Islam et al., 2017). People-centeredness emphasizes the pivotal role of individuals and their unique experiences within the healthcare context. Patient-centered care signifies the practice of tailoring healthcare services to meet the distinct needs, preferences, and expectations of each patient. This approach prioritizes values like active engagement and fostering mutually beneficial connections between healthcare providers and patients. Health improvement represents the main objective of enhancing the overall health of a population. This concept encompasses the deployment of interventions to enhance health outcomes, with the dual purpose of reducing the incidence of diseases and preventing mortality. It embodies a comprehensive healthcare approach, concentrating on the promotion of improved health outcomes for all populations.

The HSPA approach emphasizes efforts to improve health outcomes within specific health system parameters. It does not explicitly measure the impact of improving health system goals on broader social goals. Nevertheless, it recognizes the importance of these factors when analyzing health systems. For example, health systems play a critical role in promoting social cohesion, equity, and stability within society. This is achieved by alleviating the financial burdens associated with disease. Health systems support societal well-being by promoting social protection, increasing economic productivity, and improving equity. In addition, health system can impact larger societal goals. The financing structure chosen in a health care system can affect workforce mobility and thus overall economic performance. Similarly, the implementation of organizational and technological advances in healthcare can have effects on many industries and lead to higher economic performance. The correlation between health system performance and the achievement of societal goals is crucial to policymakers.

3. Blockchain in Healthcare Enhancing Business Practices

Blockchain technology, originally developed as the underlying structure for cryptocurrencies like Bitcoin (Nakamoto, 2008), has emerged as a transformative tool in various industries, including healthcare (Abbate et al., 2023). Its decentralized digital ledger system ensures that transactions are securely recorded across multiple computers, making the data immutable and transparent (O'Leary, 2017). This high level of security and transparency provided by blockchain addresses several critical challenges in healthcare, thereby enhancing business practices within the industry.

Blockchain is a distributed ledger technology (DLT) that operates on a peer-to-peer network (Bonnet and Teuteberg, 2023). Each participant, or node, in the network maintains a copy of the ledger, which is updated and validated through a consensus mechanism. Unlike traditional centralized databases, blockchain operates on a decentralized network (Lee, 2019). Each node has equal authority and maintains a copy of the entire blockchain. This decentralization eliminates single points of failure, enhancing the security and resilience of the system. Furthermore, blockchain's immutability, achieved through cryptographic hashing, ensures that once data is written to the blockchain, it cannot be altered or deleted. Each block contains a cryptographic hash of the previous block, forming a chain of blocks. Any attempt to alter a block would change its hash, breaking the chain and alerting the network to the tampering (Berdik et al., 2021; O'Leary, 2017).

There are two main types of blockchains: public and private. Public blockchains, like Bitcoin and Ethereum, are open to anyone and are characterized by their decentralized and transparent nature (Akcora et al., 2018). Anyone can join the network, participate in the consensus process, and view the transactions recorded on the blockchain. This openness provides high levels of security and trust but can also lead to scalability issues due to the high computational power required for consensus mechanisms like Proof of Work. In contrast, private blockchains are restricted to a specific group of participants. These blockchains are used within organizations or consortia where control over the network is necessary, as in the healthcare industry. Private blockchains offer faster transaction speeds and greater privacy since access is limited and controlled, making them suitable for enterprise applications (Lacity, 2022) like healthcare where data privacy and security are paramount.

One of the primary benefits of blockchain in healthcare is the secure management of patient data. Blockchain technology ensures that patient records are securely stored and immune to tampering, protecting against unauthorized access and data breaches. This secure data management is crucial in an era where health information is a prime target for cyberattacks (Akoh Atadoga et al., 2024). Furthermore, blockchain empowers patients by giving them control over their medical records. Through smart contracts, patients can grant or revoke access to their data, ensuring that sensitive information is shared only with their consent, thus maintaining patient privacy (Griggs et al., 2018).

Improving interoperability among healthcare providers is another advantage of blockchain. The sharing of medical records across different healthcare systems reduces the fragmentation of health data. This interoperability ensures that healthcare providers have access to complete and up-to-date patient information, which is essential for accurate diagnosis (Villarreal et al., 2023). Additionally, blockchain provides a standardized format for health records, facilitating better communication and data exchange between disparate systems and enhancing the overall quality of healthcare services.

The adoption of blockchain in healthcare also comes with legal considerations (Drummer and Neumann, 2020). One of the critical aspects is compliance with data protection regulations such as the General Data Protection Regulation (GDPR) in Europe and the Health Insurance Portability and Accountability Act (HIPAA) in the United States. These regulations mandate stringent measures for the protection of personal health information. Blockchain's inherent security features align well with these requirements, but challenges remain in ensuring that data is stored and processed in a manner that fully complies with legal standards. The legal status of blockchain transactions and the enforceability of smart contracts are areas of ongoing development (Geiregat, 2018). While blockchain can automate many processes, the legal system is still adapting to these new technologies. Ensuring that blockchain solutions are designed in compliance with existing laws and regulations is crucial. Legal frameworks are evolving to address these new paradigms, but healthcare organizations must stay informed and proactive in addressing these issues.

4. Enhancing HSPA Framework Implementation with Blockchain

The utilization of blockchain in healthcare can aid in implementing the HSPA framework, offering unparalleled advantages to the healthcare sector (Figure 1.4.2). The HSPA framework operates within the boundaries of the healthcare system, primarily focusing on the enhancement of health outcomes and the attainment of specified health system goals. This framework is designed to assess the efficiency of various functions within healthcare system management, all of which play an important role in ensuring the seamless operation of the healthcare system.

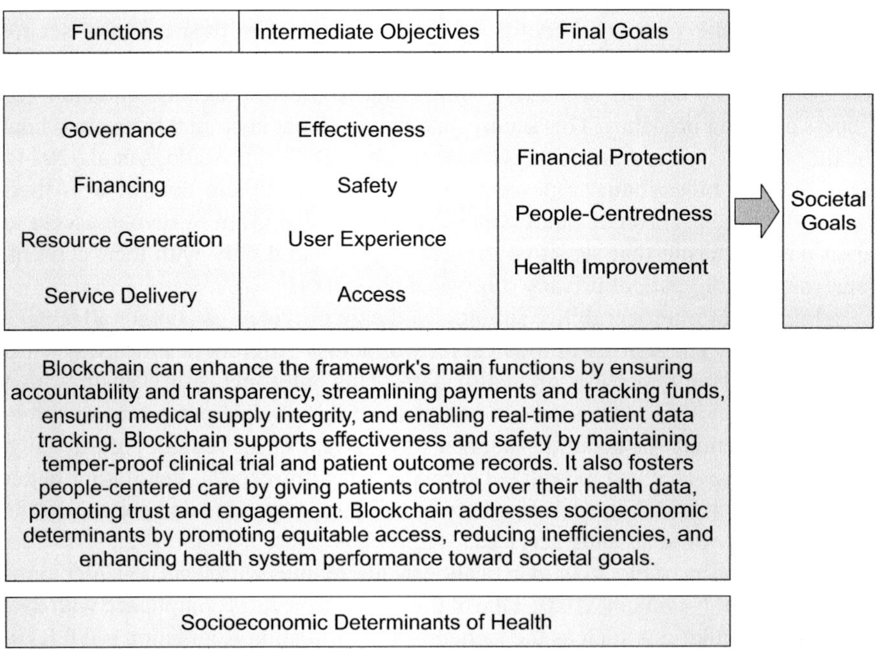

Fig. 4.2 Enhancing the HSPA framework with blockchain for better business practices

Within the healthcare system, the concept of health governance encompasses several subordinate functions with the aim of fostering collaboration across diverse sectors and utilizing legislation and regulations to advance public health goals. The central objective is to exert an influence on the determinants of health, often achieved through multifaceted collaborations with various sectors. The implementation of blockchain has the potential to establish a system that champions transparency in governance practices (Francisco and Swanson, 2018; Hellani et al., 2021). Using an immutable and decentralized ledger inherent to blockchain technology, transparency, and tamper-proofing of governance decisions and collaborative endeavors are guaranteed. This not only builds trust but also enables the creation of a comprehensive record of healthcare governance practices (Lemieux, 2017).

Healthcare financing is a key part of the healthcare system. It helps manage funding properly to deliver healthcare services, making sure everyone has fair access to healthcare. The integration of blockchain technology holds the potential to revolutionize financial transactions (Miller et al., 2023; Natanelov et al., 2022). Healthcare organizations can enhance the accuracy and transparency of various processes, including payments, provider reimbursements, and healthcare service costs. The utilization of smart contracts automates a spectrum of financial transactions, ranging from insurance claims to reimbursements, elevating system efficiency while mitigating the potential for fraudulent activities. Consequently, this practice not only engenders trust but also ensures the judicious allocation of healthcare resources.

Resource generation stands as a cornerstone in ensuring that healthcare facilities are adequately equipped with indispensable resources, encompassing medical professionals, equipment, and medications. The adoption of blockchain in supply chains presents many benefits. It guarantees the secure procurement, allocation, and maintenance of resources (Alkhudary et al., 2022; Zheng et al., 2019). By implementing blockchain technology to monitor the production and distribution of medical resources within a transparent and tamper-resistant ledger, the potential for shortages or inefficiencies within the healthcare supply chain is reduced, enhancing resource generation.

Service delivery assumes a critical role within the healthcare system, exerting an important impact on governance, financing, and resource mobilization. The incorporation of blockchain technology has the potential to enhance the efficiency of healthcare service management through real-time monitoring of service quality and patient outcomes. Blockchain technology provides a secure means of recording data related to service delivery (Al-Sumaidaee et al., 2022; Kshetri, 2017), facilitating transparent evaluation of healthcare service provision. This approach improves transparency and the capacity to make data-driven decisions aimed at enhancing the quality of services delivered.

The notion of effectiveness in healthcare, which refers to the achievement of desired goals or outcomes, aligns with the potential of blockchain technology to improve healthcare through its ability to ensure the reliable and secure transfer of healthcare data. The inherent transparency and tamper-proof properties of blockchain technology can facilitate the monitoring and management of individual patient records, treatment plans, and health outcomes (Alzahrani et al., 2023; De

Aguiar et al., 2020). Consequently, this plays a role in improving the evaluation of healthcare, which is beneficial to both individual patients and the broader population.

Improving healthcare safety with a focus on proactive measures to prevent adverse outcomes or injuries can be supported using blockchain. The immutable ledger of blockchain technology provides a secure way to store critical health data and ensure the accuracy and integrity of patient records and treatment protocols. Blockchain technology can improve the integration of health data and mitigate risk and protect patient welfare (Hölbl et al., 2018).

The concept of user experience focuses on patient experience with healthcare services. The use of blockchain technology has the potential to improve the exchange of health data between healthcare providers. It can also promote a collaborative and patient-centered approach to healthcare. Patients can be more involved in the decision-making process regarding their healthcare, resulting in higher satisfaction. This can be achieved by integrating non-fungible tokens (NFTs), a new trend based on blockchain technology (Al-Sumaidaee, Alkhudary, and Zilic, 2023; Lacity and Treiblmaier, 2022), into the healthcare system.

Ensuring access to healthcare services is a primary goal of healthcare systems, along with financial protection. The use of blockchain technology has the potential to facilitate the optimization of health insurance administrative processes and reduce the burden of administrative costs. The use of blockchain technology can improve the accessibility and availability of healthcare services. This is due to the transparency and efficiency provided by blockchain (Wamba and Queiroz, 2020), which in turn ensures financial protection mechanisms.

To finish, blockchain can engender broader societal implications (Chaudhuri et al., 2021). The amalgamation of healthcare and societal goals has the potential to bolster macroeconomic development, foster social cohesion, and enhance the overall well-being of society. The utilization of blockchain within healthcare systems contributes to public health and economic productivity. The inherent characteristics of openness and trust-building intrinsic to this technology can fortify social protection, equity, and stability within society, thereby aligning with overarching societal goals (Swan, 2015). Furthermore, the integration of blockchain technology in healthcare holds the promise of increasing workforce mobility and contributing to improved economic performance.

5. Future Research

This section delves into the application of benchmarking as a tool for assessing the influence of blockchain technology on health systems. Utilizing benchmarking techniques, researchers can evaluate the extent to which blockchain implementation affects various facets of healthcare, including data security, and interoperability. Benchmarking has evolved into a widely adopted strategy for enhancing organizational performance in various industries (Pradhan et al., 2022). This approach is rooted in the belief that gaining insights from successful business practices is more advantageous than developing new solutions from the ground up.

The concept of benchmarking entails an ongoing process of measuring and appraising products, services, and processes with the objective of identifying optimal approaches and strategies (Åhrén and Parida, 2009; Al-Tarawneh, 2014). The adoption of benchmarking principles has experienced significant growth across multiple sectors, driven by the aspiration to achieve peak performance and foster innovation. Incorporating benchmarking methodologies into the healthcare industry provides an advantageous framework for scrutinizing the impact of blockchain technology on the quality and performance of healthcare systems.

A fundamental element of benchmarking in the healthcare industry is competitive benchmarking, which assesses the operational efficiency of healthcare providers. The benchmarking method can be applied to conduct a comparative analysis of healthcare systems sharing similar characteristics, such as their public or non-public status, the scope of medical services they offer, their territorial reach, and their size (Ozcan, 2014). Future research can study the degree to which blockchain enhances the competitiveness of healthcare providers. It can also evaluate the feasibility of blockchain in healthcare systems by conducting in-depth case studies of hospitals that have successfully implemented decentralized applications.

Furthermore, in addition to competitive benchmarking, the incorporation of functional benchmarking is crucial for assessing the broader ramifications of blockchain in healthcare. This concept extends beyond direct competition and involves comparing healthcare providers with differing profiles but overlapping operational domains. Researchers should delve into the potential impact of blockchain on shared services, such as resource and equipment management, laboratory operations, and hospital pharmacy management. This approach allows for a more comprehensive understanding of how blockchain enhances operational performance and optimizes resource allocation in healthcare (Agbo et al., 2019; Akoh Atadoga et al., 2024).

In the context of specialized healthcare organizations consisting of multiple departments offering similar or multispecialty medical services, internal benchmarking becomes a good practice (Agarwal et al., 2016; Dugas et al., 2008; García-Lorenzo et al., 2023). Researchers have explored the impact of blockchain on various domains, encompassing quality improvement, patient safety, healthcare supply chain management, capital project performance, financial metrics, and the efficiency of primary care centers (Ang et al., 2015; Cimasi, 2004). Further research should be undertaken to examine the impact of blockchain technology on the financial aspects of medical services. By conducting a comparative analysis of key indicators, including medical staff-to-patient ratios, bed occupancy rates, and person-days, researchers can gain insights into how blockchain adoption improves the delivery of healthcare services and the management of resources.

Conclusion

The healthcare industry operates in a dynamic environment characterized by constant regulatory changes, budget constraints, and technological advances, all of which impact its performance and quality. The dynamic healthcare environment faces several challenges in managing resources, complying with regulatory requirements, and ensuring equitable access to services while striving for cost efficiency. However, achieving these goals is complicated by information asymmetries and demographic changes, necessitating the discovery of new solutions. The emergence of blockchain is being viewed as a new information system that can reshape the structure and operation of the healthcare industry. It can enhance transparency and security by addressing issues related to governance, finance, resource generation, and service delivery. The proposed system resolves data sharing and security concerns and fosters trust among multiple stakeholders. It also promotes the implementation of patient-centered care and increases financial security. Integrating blockchain into the HSPA framework not only enhances the healthcare operations but also aligns with broader societal goals. This improvement can impact social inclusion, economic development, and overall prosperity.

Linking blockchain to better business practices in healthcare reveals several advantages. By streamlining administrative processes and reducing redundancies, blockchain improves resource management, enabling healthcare providers to avoid duplication of tests and procedures, thus reducing costs and improving patient outcomes. Blockchain's secure and transparent nature ensures accurate recording of transactions, simplifying regulatory compliance. Moreover, it facilitates equitable access to services for tracking resource allocation, ensuring that underserved populations receive appropriate care and helping to identify gaps in service delivery. Blockchain also supports continuous, up-to-date access to patient records for ongoing treatment and emergency situations. This fosters technical competence among healthcare providers, allowing them to make better decisions. Additionally, blockchain reduces information asymmetries by ensuring that all stakeholders have access to the same, up-to-date information, thereby building trust and enhancing the quality of healthcare services. As populations age and demographic shifts occur, blockchain aids in the development of innovative solutions, such as personalized medicine and telehealth services.

Furthermore, blockchain contributes to social inclusion by ensuring all individuals have access to quality healthcare services. The efficiencies gained through blockchain can lead to cost savings and improved economic performance within the healthcare sector, fostering broader economic development. Finally, blockchain facilitates the secure and verifiable transfer of professional credentials and employment history across borders, promoting labor mobility and enabling healthcare professionals to work in different regions or countries without the hassle of re-verification. This mobility is essential for addressing workforce shortages and ensuring skilled professionals can respond to areas of need quickly. Consequently, blockchain offers a comprehensive approach to better business practices in the healthcare industry.

Despite its potential, the implementation of blockchain in healthcare also presents several limitations. One major challenge is the scalability of blockchain systems. Blockchain can suffer from slow transaction speeds due to the high computational power required for consensus mechanisms. Additionally, there are considerable legal and regulatory hurdles to overcome. Ensuring compliance with data protection regulations such as GDPR and HIPAA can be complex, and the legal status and enforceability of smart contracts are still evolving areas. There are also concerns about the interoperability of blockchain with existing healthcare information systems, which could require additional adjustments and investments. Finally, the utilization of blockchain necessitates initial costs, which may be prohibitive for some healthcare organizations. There is also a need for widespread education and training among healthcare providers to implement and utilize blockchain technology.

References

Abbate, S., Centobelli, P., Cerchione, R., Oropallo, E. and Riccio, E. (2023). Blockchain Technology for Embracing Healthcare 4.0. *IEEE Transactions on Engineering Management*, *70*(8), 2998–3009. https://doi.org/10.1109/TEM.2022.3212007.

Agarwal, R., Green, R., Agarwal, N. and Randhawa, K. (2016). Benchmarking management practices in Australian public healthcare. *Journal of Health Organization and Management*, *30*(1), 31–56. https://doi.org/10.1108/JHOM-07-2013-0143.

Agbo, C.C., Mahmoud, Q.H. and Eklund, J.M. (2019). Blockchain Technology in Healthcare: A Systematic Review. *Healthcare*, *7*(2), 56. https://doi.org/10.3390/healthcare7020056.

Åhrén, T. and Parida, A. (2009). Maintenance performance indicators (MPIs) for benchmarking the railway infrastructure: A case study. *Benchmarking: An International Journal*, *16*(2), 247–58. https://doi.org/10.1108/14635770910948240.

Akcora, C.G., Dixon, M. F., Gel, Y.R. and Kantarcioglu, M. (2018). Bitcoin risk modeling with blockchain graphs. *Economics Letters*, *173*, 138–42. https://doi.org/10.1016/j.econlet.2018.07.039.

Akoh Atadoga, Oluwafunmi Adijat Elufioye, Toritsemogba Tosanbami Omaghomi, Opeoluwa Akomolafe, Ifeoma Pamela Odilibe, & Oluwaseyi Rita Owolabi. (2024). Blockchain in healthcare: A comprehensive review of applications and security concerns. *International Journal of Science and Research Archive*, *11*(1), 1605–13. https://doi.org/10.30574/ijsra.2024.11.1.0244.

Alkhudary, R. and Féniès, P. (2022). Blockchain and Trust in Supply Chain Management: A Conceptual Framework. *IFAC-PapersOnLine*, *55*(10), 2402–2406. https://doi.org/10.1016/j.ifacol.2022.10.068.

Alkhudary, R., Queiroz, M.M. and Féniès, P. (2022). Mitigating the risk of specific supply chain disruptions through blockchain technology. *Supply Chain Forum: An International Journal*, 1–11. https://doi.org/10.1080/16258312.2022.2090273.

Allen-Duck, A., Robinson, J.C. and Stewart, M.W. (2017). Healthcare Quality: A Concept Analysis. *Nursing Forum*, *52*(4), 377–86. https://doi.org/10.1111/nuf.12207.

Alonazi, W.B. and Thomas, S.A. (2014). Quality of Care and Quality of Life: Convergence or Divergence? *Health Services Insights*, *7*, HSI.S13283. https://doi.org/10.4137/HSI.S13283.

Al-Sumaidaee, G., Alexandridis, A., Alkhudary, R. and Zilic, Z. (2022). A Technical Assessment of Blockchain in Healthcare with a Focus on Big Data. 2022 IEEE International Conference on Big Data (Big Data), 2467–72. https://doi.org/10.1109/BigData55660.2022.10020962.

Al-Sumaidaee, G., Alkhudary, R. and Zilic, Z. (2023). Non-Fungible Tokens (NFTs) as a Means for Blockchain Networks Integration in Healthcare. 2023 5th Conference on Blockchain Research

& Applications for Innovative Networks and Services (BRAINS), 1–2. https://doi.org/10.1109/BRAINS59668.2023.10316782.

Al-Sumaidaee, G., Alkhudary, R., Zilic, Z. and Féniès, P. (2021). A blueprint towards an integrated healthcare information system through blockchain technology. HEALTHINFO 2021, The Sixth International Conference on Informatics and Assistive Technologies for Health-Care, Medical Support and Wellbeing.

Al-Sumaidaee, G., Alkhudary, R., Zilic, Z. and Féniès, P. (2023). Configuring Blockchain Architectures and Consensus Mechanisms: The Healthcare Supply Chain as a Use Case. In A. Bouras, I. Khalil, & B. Aouni (Eds.), *Blockchain Driven Supply Chains and Enterprise Information Systems* (pp. 135–50). Springer International Publishing. https://doi.org/10.1007/978-3-030-96154-1_7.

Al-Tarawneh, H. (2014). The Utilization of Benchmarking in the Jordanian Banking Sector. *Journal of Management Research*, *6*(3), 49. https://doi.org/10.5296/jmr.v6i3.5299.

Alzahrani, S., Daim, T. and Choo, K.-K. R. (2023). Assessment of the Blockchain Technology Adoption for the Management of the Electronic Health Record Systems. *IEEE Transactions on Engineering Management*, *70*(8), 2846–63. https://doi.org/10.1109/TEM.2022.3158185.

Ang, D., McKenney, M., Norwood, S., Kurek, S., Kimbrell, B., Liu, H., Ziglar, M. and Hurst, J. (2015). Benchmarking statewide trauma mortality using Agency for Healthcare Research and Quality's patient safety indicators. *Journal of Surgical Research*, *198*(1), 34–40. https://doi.org/10.1016/j.jss.2015.05.053.

Aydin, A. (2022). Benchmarking healthcare systems of OECD countries: A DEA-based Malmquist Productivity Index Approach. *Alphanumeric Journal*, *10*(1), 25–40. https://doi.org/10.17093/alphanumeric.1057559.

Barbazza, E., Kringos, D., Kruse, I., Klazinga, N.S. and Tello, J.E. (2019). Creating performance intelligence for primary health care strengthening in Europe. *BMC Health Services Research*, *19*(1), 1006. https://doi.org/10.1186/s12913-019-4853-z.

Beaulieu, M., Bentahar, O., Benzidia, S. and Gunasekaran, A. (2023). Digitalization Initiatives of Home Care Medical Supply Chain: A Case-Study-Based Approach. *IEEE Transactions on Engineering Management*, 1–14. https://doi.org/10.1109/TEM.2023.3265624.

Benzidia, S., Ageron, B., Bentahar, O. and Husson, J. (2019). Investigating automation and AGV in healthcare logistics: A case study based approach. *International Journal of Logistics Research and Applications*, *22*(3), 273–93. https://doi.org/10.1080/13675567.2018.1518414.

Berdik, D., Otoum, S., Schmidt, N., Porter, D. and Jararweh, Y. (2021). A Survey on Blockchain for Information Systems Management and Security. *Information Processing & Management*, *58*(1), 102397. https://doi.org/10.1016/j.ipm.2020.102397.

Bernal-Delgado, E., Christiansen, T., Bloor, K., Mateus, C., Yazbeck, A.M., Munck, J. and Bremner, J. (2015). ECHO: Health care performance assessment in several European health systems. *European Journal of Public Health*, *25*(suppl_1), 3–7. https://doi.org/10.1093/eurpub/cku219.

Bonnet, S. and Teuteberg, F. (2023). Impact of blockchain and distributed ledger technology for the management of the intellectual property life cycle: A multiple case study analysis. *Computers in Industry*, *144*, 103789. https://doi.org/10.1016/j.compind.2022.103789.

Chaudhuri, A., Bhatia, M.S., Kayikci, Y., Fernandes, K.J. and Fosso-Wamba, S. (2021). Improving social sustainability and reducing supply chain risks through blockchain implementation: Role of outcome and behavioural mechanisms. Annals of Operations Research. https://doi.org/10.1007/s10479-021-04307-6.

Cimasi, R.J. (2004, November 1). Financial Benchmarking in the Health Care Industry Part II. https://www.cpaleadership.com/public/257.cfm.

De Aguiar, E.J., Faiçal, B.S., Krishnamachari, B. and Ueyama, J. (2020). A Survey of Blockchain-Based Strategies for Healthcare. *ACM Computing Surveys*, *53*(2), 1–27. https://doi.org/10.1145/3376915.

Drummer, D. and Neumann, D. (2020). Is code law? Current legal and technical adoption issues and remedies for blockchain-enabled smart contracts. *Journal of Information Technology*, *35*(4), 337–60. https://doi.org/10.1177/0268396220924669.

Dugas, M., Eckholt, M. and Bunzemeier, H. (2008). Benchmarking of hospital information systems: Monitoring of discharge letters and scheduling can reveal heterogeneities and time trends. *BMC Medical Informatics and Decision Making*, *8*(1), 15. https://doi.org/10.1186/1472-6947-8-15.

Francisco, K. and Swanson, D. (2018). The Supply Chain Has No Clothes: Technology Adoption of Blockchain for Supply Chain Transparency. *Logistics*, *2*(1), 2. https://doi.org/10.3390/logistics2010002.

Fuchs, V.R. (2018). *Health Economics and Policy: Selected Writings by Victor Fuchs* (1st edition). World Scientific Publishing Co Pte Ltd..

García-Lorenzo, B., Gorostiza, A., Alayo, I., Castelo Zas, S., Cobos Baena, P., Gallego Camiña, I., Izaguirre Narbaiza, B., Mallabiabarrena, G., Ustarroz-Aguirre, I., Rigabert, A., Balzi, W., Maltoni, R., Massa, I., Álvarez López, I., Arévalo Lobera, S., Esteban, M., Fernández Calleja, M., Gómez Mediavilla, J., Fernández, M., … Barros, E. (2023). European value-based healthcare benchmarking: Moving from theory to practice. *European Journal of Public Health*, ckad181. https://doi.org/10.1093/eurpub/ckad181.

Geiregat, S. (2018). Cryptocurrencies are (smart) contracts. *Computer Law & Security Review*, *34*(5), 1144–49. https://doi.org/10.1016/j.clsr.2018.05.030.

Griggs, K.N., Ossipova, O., Kohlios, C.P., Baccarini, A.N., Howson, E.A. and Hayajneh, T. (2018). Healthcare Blockchain System Using Smart Contracts for Secure Automated Remote Patient Monitoring. *Journal of Medical Systems*, *42*(7), 130. https://doi.org/10.1007/s10916-018-0982-x.

Hargett, C., Doty, J., Hauck, J., Webb, A., Cook, S., Tsipis, N., Neumann, J., Andolsek, K. and Taylor, D. (2017). Developing a model for effective leadership in healthcare: A concept mapping approach. *Journal of Healthcare Leadership*, *Volume 9*, 69–78. https://doi.org/10.2147/JHL.S141664.

Hellani, H., Sliman, L., Samhat, A.E. and Exposito, E. (2021). On Blockchain Integration with Supply Chain: Overview on Data Transparency. *Logistics*, *5*(3), 46. https://doi.org/10.3390/logistics5030046.

Hölbl, M., Kompara, M., Kamišalić, A. and Nemec Zlatolas, L. (2018). A Systematic Review of the Use of Blockchain in Healthcare. *Symmetry*, *10*(10), 470. https://doi.org/10.3390/sym10100470.

Islam, Md. R., Rahman, Md. S., Islam, Z., Nurs, C.Z.B., Sultana, P. and Rahman, Md. M. (2017). Inequalities in financial risk protection in Bangladesh: An assessment of universal health coverage. *International Journal for Equity in Health*, *16*(1), 59. https://doi.org/10.1186/s12939-017-0556-4.

KirungaTashobya, C., Ssengooba, F., Nabyonga-Orem, J., Bataringaya, J., Macq, J., Marchal, B., Musila, T. and Criel, B. (2018). A critique of the Uganda district league table using a normative health system performance assessment framework. *BMC Health Services Research*, *18*(1), 355. https://doi.org/10.1186/s12913-018-3126-6.

Kludacz-Alessandri, M., Walczak, R., Hawrysz, L. and Korneta, P. (2021). The Quality of Medical Care in the Conditions of the COVID-19 Pandemic, with Particular Emphasis on the Access to Primary Healthcare and the Effectiveness of Treatment in Poland. *Journal of Clinical Medicine*, *10*(16), 3502. https://doi.org/10.3390/jcm10163502.

Kshetri, N. (2017). Blockchain's roles in strengthening cybersecurity and protecting privacy. *Telecommunications Policy*, *41*(10), 1027–38. https://doi.org/10.1016/j.telpol.2017.09.003.

Lacity, M.C. (2022). Blockchain: From Bitcoin to the Internet of Value and beyond. *Journal of Information Technology*, 026839622210863. https://doi.org/10.1177/02683962221086300.

Lacity, M.C. and Treiblmaier, H. (Eds.). (2022). *Blockchains and the Token Economy: Theory and Practice*. Springer International Publishing. https://doi.org/10.1007/978-3-030-95108-5.

Lee, J.Y. (2019). A decentralized token economy: How blockchain and cryptocurrency can revolutionize business. *Business Horizons*, *62*(6), 773–84. https://doi.org/10.1016/j.bushor.2019.08.003.

Lemieux, V.L. (2017). A typology of blockchain recordkeeping solutions and some reflections on their implications for the future of archival preservation. 2017 IEEE International Conference on Big Data (Big Data), 2271–78. https://doi.org/10.1109/BigData.2017.8258180.

Miller, T., Cao, S., Foth, M., Boyen, X. and Powell, W. (2023). An asset-backed decentralised finance instrument for food supply chains – A case study from the livestock export industry. *Computers in Industry*, *147*, 103863. https://doi.org/10.1016/j.compind.2023.103863.

Nakamoto, S. (2008). *Bitcoin: A Peer-to-Peer Electronic Cash System*. https://bitcoin.org/bitcoin.pdf. https://bitcoin.org/bitcoin.pdf.

Natanelov, V., Cao, S., Foth, M. and Dulleck, U. (2022). Blockchain smart contracts for supply chain finance: Mapping the innovation potential in Australia-China beef supply chains. *Journal of Industrial Information Integration*, *30*, 100389. https://doi.org/10.1016/j.jii.2022.100389.

Noto, G., Corazza, I., Kļaviņa, K., Lepiksone, J. and Nuti, S. (2019). Health system performance assessment in small countries: The case study of Latvia. *The International Journal of Health Planning and Management*, *34*(4), 1408–22. https://doi.org/10.1002/hpm.2803.

Ntr, A., M, R.A. and C, L.P. (2019). The Literature Review of the Governance Frameworks in Health System. *Journal of Public Administration and Governance*, *9*(3), 252. https://doi.org/10.5296/jpag.v9i3.15535.

O'Leary, D.E. (2017). Configuring blockchain architectures for transaction information in blockchain consortiums: The case of accounting and supply chain systems. *Intelligent Systems in Accounting, Finance and Management*, *24*(4), 138–47. https://doi.org/10.1002/isaf.1417.

Ozcan, Y.A. (2014). *Health Care Benchmarking and Performance Evaluation: An Assessment using Data Envelopment Analysis (DEA)* (Vol. 210). Springer US. https://doi.org/10.1007/978-1-4899-7472-3.

Papanicolas, I., Rajan, D., Karanikolos, M., Soucat, A. and Figueras, J. (2022). *Health system performance assessment a framework for policy analysis*. World Health Organization.

Pradhan, N.R., Singh, A.P., Verma, S., Kavita, Kaur, N., Roy, D.S., Shafi, J., Wozniak, M. and Ijaz, M.F. (2022). A Novel Blockchain-Based Healthcare System Design and Performance Benchmarking on a Multi-Hosted Testbed. *Sensors*, *22*(9), 3449. https://doi.org/10.3390/s22093449.

Rejeb, A., Rejeb, K., Treiblmaier, H., Appolloni, A., Alghamdi, S., Alhasawi, Y. and Iranmanesh, M. (2023). The Internet of Things (IoT) in healthcare: Taking stock and moving forward. *Internet of Things*, *22*, 100721. https://doi.org/10.1016/j.iot.2023.100721.

Sanford, N., Lavelle, M., Markiewicz, O., Reedy, G., Rafferty, D.A.M., Darzi, L.A. and Anderson, J. E. (2024). Decoding healthcare teamwork: A typology of hospital teams. *Journal of Interprofessional Care*, *38*(4), 602–611. https://doi.org/10.1080/13561820.2024.2343835.

Shortell, S.M. (1989). *Health Care Management: A Text in Organization Theory & Behavior*. Delmar.

Singh, M.D. and Barakoti, B. (2023). The Impact of Social Health Insurance on Household Catastrophic Healthcare Expenditures Associated with Chronic Diseases in Sundarharaincha Municipality, Morang. *Southwestern Research Journal*, *1*(1), 39–48. https://doi.org/10.3126/srj.v1i1.62263.

Smith, P.C., Anell, A., Busse, R., Crivelli, L., Healy, J., Lindahl, A.K., Westert, G. and Kene, T. (2012). Leadership and governance in seven developed health systems. *Health Policy*, *106*(1), 37–49. https://doi.org/10.1016/j.healthpol.2011.12.009.

Swan, M. (2015). Blockchain: Blueprint for a new economy (First edition). O'Reilly.

Villarreal, E.R.D., Garcia-Alonso, J., Moguel, E. and Alegria, J.A.H. (2023). Blockchain for Healthcare Management Systems: A Survey on Interoperability and Security. *IEEE Access*, *11*, 5629–52. https://doi.org/10.1109/ACCESS.2023.3236505.

Wamba, S.F. and Queiroz, M.M. (2020). Blockchain in the operations and supply chain management: Benefits, challenges and future research opportunities. *International Journal of Information Management*, *52*, 102064. https://doi.org/10.1016/j.ijinfomgt.2019.102064.

Zheng, K., Zhang, Z. (Justin), Chen, Y. and Wu, J. (2019). Blockchain adoption for information sharing: Risk decision-making in spacecraft supply chain. Enterprise Information Systems, 1–22. https://doi.org/10.1080/17517575.2019.1669831.

5 Money as a Design Material Towards the Empowerment of Citizens

Ryan Bowler[1], Chris Speed[2*] and Geoffrey Goodell[3]

1. Introduction

Throughout history, currency has taken many forms, from shells and metal coins and more recently, non-regulated cryptocurrencies. As governments explore their own versions of digital currency, such as Central Bank Digital Currencies (CBDC), design has an increasingly important role to play in preserving the many values associated with money. This chapter invites designers to reflect on their role in shaping or maintaining the socio-cultural values and practices associated with money, both in established forms such as physical cash and new forms including CBDCs.

Money represents more than just a numeric value to people; it embodies trust, and social morals, and offers flexible ways to transact through its tangible and intangible properties. In doing so, money binds social relationships (Halloluwa et al., 2018) and plays an invaluable role in people's daily routines (Vines et al., 2012). Money binds many parts of society together (Ingham, 1996), and takes on the most unusual forms within our lives as it mutates from cash into lottery tickets, written cheques, equivalent value in supermarket and airline loyalty schemes, or the simple 'tap' of a digital payment from a smart phone. The form that money takes is both highly creative but is never without some level of complexity that places the citizen within increasing material, digital and social entanglements (Speed et al., 2022). Maurer (2006) underlines these complexities by constructing an anthropology through which we see how money has become a knotty relationship between "'real' economic value and [the] 'insubstantial' fictions" that government-backed currencies (so-called fiat money), stock markets and organisations have produced. Whilst societies have experimented with the use of cryptocurrencies as a novel form of money, it is only now that we are seeing governments invest significantly in their own entirely digital forms of money—commonly known as Central Bank Digital Currencies (CBDC). Whilst China has been seen at the forefront of CBDC development with its digital yuan project, also known as the Digital Currency Electronic Payment (DCEP) system, Sweden, Norway, and the

[1] School of Informatics, The University of Edinburgh, UK
[2] RMIT University, Melbourne, Australia
[3] Department of Computer Science, University College London, UK
Email: r.bowler@sms.ed.ac.uk[1]; g.goodell@ucl.ac.uk[3]
* Corresponding author: chris.speed@rmit.edu.au

UK have all initiated research into the viability of such schemes. Meanwhile, in October of 2020, the Bahamas launched its 'Sand Dollar,' an entirely digital version of the Bahamian dollar. However, if the current trend away from the representation of value in the form of physical cash continues, there is a real risk that people who rely on material forms of money will be excluded, and their values and prior practices impacted. It is important to note that whilst the rise of CBDC signals an adoption of blockchain technology by governments, the use of digital technology to support economic transactions are not new. M-Pesa, is a digital mobile money service that was launched in 2007 by Vodafone and Safaricom, the largest mobile network operators in Kenya, and enables users to deposit money into an account stored on their mobile phones, to send balances across SMS messages to other users, and to redeem deposits for regular money (Centellegher et al., 2018; Roitman, 2023).

As we look towards the fast-approaching horizon of CBDC, our findings suggest that the speed at which it is being discussed is moving too quickly and there are still many questions that need to be answered. We report how CBDC offers new ways to consider how values can be associated with value, but that it must also echo certain present monetary values, such as cash's anonymity that can help avoid discrimination. Furthermore, understanding how CBDC will be designed can allow us to evaluate which user perspectives will be key in shaping both its tangible and interaction design. These are some of the considerations we report on, based on a participatory workshop and semi-structured interviews with partners who have current knowledge or involvement in the field of digital banking. Much of what is known to date on digital currencies is theoretical, speculation or technical, and their social, political, economic and global impact is yet to be realised. This chapter serves as a starting point for designers to consider their role in shaping or maintaining the values associated with money, across both established forms such as cash, and new forms like Central Bank Digital Currency. The chapter findings offer further contributions to the questions posed in the book about how blockchain technologies are an enabler for positive impact in industry, community, and the planet.

The chapter continues with a section that provides background to the tangible aspects of money, its historical forms, and the underlying concept of debt, before introducing CBDC, its potential impact on the financial landscape, and the need to consider socio-cultural values in its design. The paper then introduces the methods used, including the participatory workshops and semi-structured interviews, both of which were informed by the use of a comic strip to illustrate potential CBDC scenarios. Findings from the workshops and interviews are explored before conclusions are drawn, highlighting the study's limitations and considering possibilities for future research directions and design considerations for CBDC.

2. Money, Money, Money

Money in the form of cash can be thought of from an engineering perspective. It is a material or fabric that is transformed into a product (Sousa et al., 2021).

Traditionally, cash is made of significant metal or paper-like material that is crafted into a tangible asset that can be exchanged or swapped (Ingham, 2013). Some have claimed that money has been essential since the beginning of humanity in various forms, from shells to coins, as it represents trust and the promise to repay or purchase (Arvidsson, 2019).

Underpinning these material forms of cash is debt, classed as a fundamental element that ignites the essential design principles of money (Desan, 2021).

"Money in all forms exhibit three principal design components: they are made of debt; that debt is specifically fashioned to create liquidity; and the debt (or credit) medium that results comes with a pledge of value. Those design components recur across major money types, including medieval coins, modern base (high-powered) money, and commercial bank money" (Desan, 2021).

The concept of debt significantly shapes the materials of money, resulting in the creation of assets such as cash, stocks, bonds, and real estate. Each of these assets has its own value and methods of conversion into cash. It can be easy to consider money as only these tangible materials, which we design for and around. However, with digitisation, the nature of money is changing.

The once tangible material for storing value is increasingly becoming an immaterial store of value in digital form (Peneder, 2022). This means that physical forms of money, such as coins and banknotes, are being replaced by digital representations that can be stored and transferred electronically. As money becomes increasingly digital, economists race to question what money is in its digital form, while software engineers continue, at pace, to digitise the financial landscape (Peneder, 2022).

A CBDC is a digital form of fiat currency money that is issued by a central bank. It is designed to be a digital representation of the country's physical currency and unlike cryptocurrencies such Bitcoin or Ethereum, CBDC is backed by the government and is legal tender. However, like Bitcoin and Etheruem, CBDCs use secure, transparent ledgers, which often use blockchain technology to create an immutable record of all transactions.

CBDCs can be considered as foundational innovations that provide the infrastructure of financial systems (Schindler, 2017). Not only could this technology have major financial implications, but equally, vital social, political and economic impacts unseen for decades. Its functions are different from that of current digital methods of exchange. They are proposed to be issued by the central bank. Reminiscent of how tactile cash is printed, governed, and distributed to populations, CBDC is distributed in the same manner, but in digital form, and with the potential support of third parties. In the UK, this digital 'coin' could be known as the digital sterling or digital pound. Due to its potential benefits and impact, CBDC has garnered significant attention and popularity in recent years.

While countries like China, and Central Europe are rapidly adopting CBDCs as a complementary government backed currency, others such as Sweden are on track to becoming an entirely cashless society, due to many social and technological factors (Arvidsson, 2018, 2019). Digital money, such as CBDC, is poised to shape a new financial landscape, while the current landscape is already undergoing changes. The UK has begun to investigate the concept of a CBDC through a joint effort by the Bank of England and HM Treasury resulting in a Phase 1 consultation paper (HM Treasury and Bank of England, 2023). CBDC in this report is to be seen as a new form of payment that is not aimed at replacing cash or other payment systems, but that it might. While this is welcoming, established forms of money, like cash, are already in decline (HM Treasury and Bank of England, 2023). This reduction is not yet due to CBDC but other factors, including the COVID-19 pandemic (UK Finance, 2021), that have hastened the pace of adopting contact-less forms of payment (Arvidsson, 2018; Iqbal and Campbell, 2020; Wiberg, 2020).

The physical form of cash remains an essential part of daily life for many people in the UK (Bank of England, 2021). But the infrastructures that support access to cash—such as Automatic Teller Machines (ATMs)—are diminishing (Ceeney, 2019). Current government policies do not adequately address the need to protect established mechanisms for accessing cash, including bank withdrawals or aim to legalise its acceptance. The decline in cash usage and the lack of measures to protect it underscore the need to understand how public values and expectations, with respect to how individuals engage with the economy, may change in a world with CBDC. Given that the first 'product' from blockchain was a form of currency, and that Bitcoin remains synonymous with the technology in the minds of the public, the academic community has to take responsibility for attending to whether the future of money in one form (CBDC) is good for society. There are many reasons for how blockchain offers an opportunity to rethink how value is created in distributed and decentralised forms, but these intellectual and technical 'investments' in the technology should not be isolated from the implications of potentially excluding members of the community, specifically those who prefer to use cash.

Failing to consider not only the economic but also the social role of cash as a means of payment could result in the creation of a CBDC that does not address the myriad of socio-cultural concerns, practices, needs or values surrounding money. Accounting for the social and political role of money in society as the materials of money change is particularly important because, as Swierstra (2013) posits, introducing technological advancements can change our perceptions of morals and values, with the potential for both negative and positive consequences. The moral outcomes that arise from introducing new technology depend on whether diverse members of an economy are considered and have input in its development (Swierstra, 2013). Research across human-computer interaction has shone light on particular ways of reconsidering 'non users' of technology in order to balance the assumption that people who use technology are 'good users' and those who do not are somehow problematic (Satchell and Dourish, 2009; Baumer et al., 2014; Selwyn, 2003). Describing this assumption as a 'morality of adoption' Satchell and Dourish (2009) highlight a responsibility for researchers to recognise digital media (such as digital money) as cultural objects, and that technology (such as blockchain) as a cultural phenomenon. In doing so, this responsibility prompts us to consider the individuals who could be impacted by a blockchain technology when it is manifest as CBDC, and given their current values and concerns, whether their perceptions, values and morals are represented in a system wherein CBDC is a dominant mode of payment.

There is still much we have yet to discover. However, we know, as previously discussed, that as CBDC is being weighed by policymakers, digital inequality remains a pressing issue and certain requirements for making digital finance accessible are being overlooked. Money is deeply intertwined with all areas of society, and its social value exceeds its economic worth. Established forms of money are steadily being displaced and technologies can inform and instill new practices, values and morals but this takes time. These factors make the design of novel money a grand challenge that requires attentiveness, patience, and criticality. Without such an approach, there is a risk of negatively impacting millions of people.

3. Participants, Methods and Findings

To better understand the role of interaction design in the CBDC landscape, we reached out to several user groups and teams involved with CBDC in various capacities. These included UK public bodies, Web 3.0 specialists, experts in commercial banking, and digital payment industry professionals. Participants were invited to take part in a small study that was foregrounded by a design workshop that elicited issues surrounding CBDC, that in turn informed semi-structured interviews. Despite the small sample size of four subsequent participants, their knowledge and expertise provided rich and valuable insights. These insights encourage designers to explore current established forms of money and the future CBDC landscape.

The comic strip that was used in the participatory workshop illustrated how a citizen called Alice might use CBDC to pay for a coffee and croissant in Bob's coffee shop (Figure 5.1). The comic was developed through conversations between the design and technical teams on the project, with advice from industrial teams. Design meetings with members were facilitated by us in such a way as to best co-design both a human-centred experience, and also provide advice on the most likely way in which software and hardware technologies might interface with national and retail banks to support a transaction. A comic illustrator was present in all meetings and was able to offer versions of the sequence of events, so that the teams could adapt and iterate the sequence of events, and the assemblage of technologies involved (Figure 5.2).

Two participatory workshops were organised in Edinburgh and London involving 10-20 participants at each event, with representation from the Bank of England, The British Standards Institution, NatWest Banking Group, Everis (a financial consulting firm), TODAQ (a micro-payment service provider), and the UK tax service HMRC. In both workshops, the comic strip was attached to the wall. Following an introduction to the aims of the project and a verbal overview

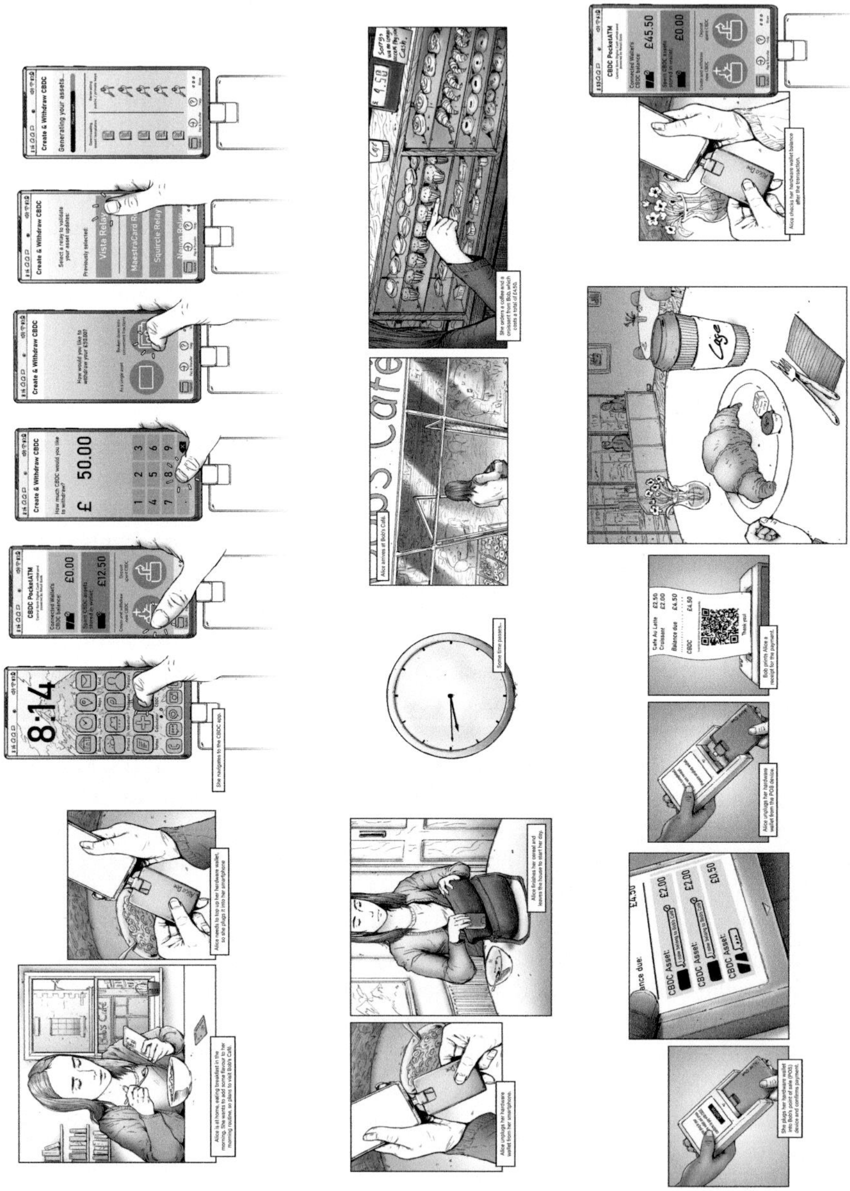

Fig. 5.1 The illustrated comic that describes how a citizen called Alice might use CBDC to pay for a coffee and croissant in Bob's coffee shop using a CBDC. Image by the authors.

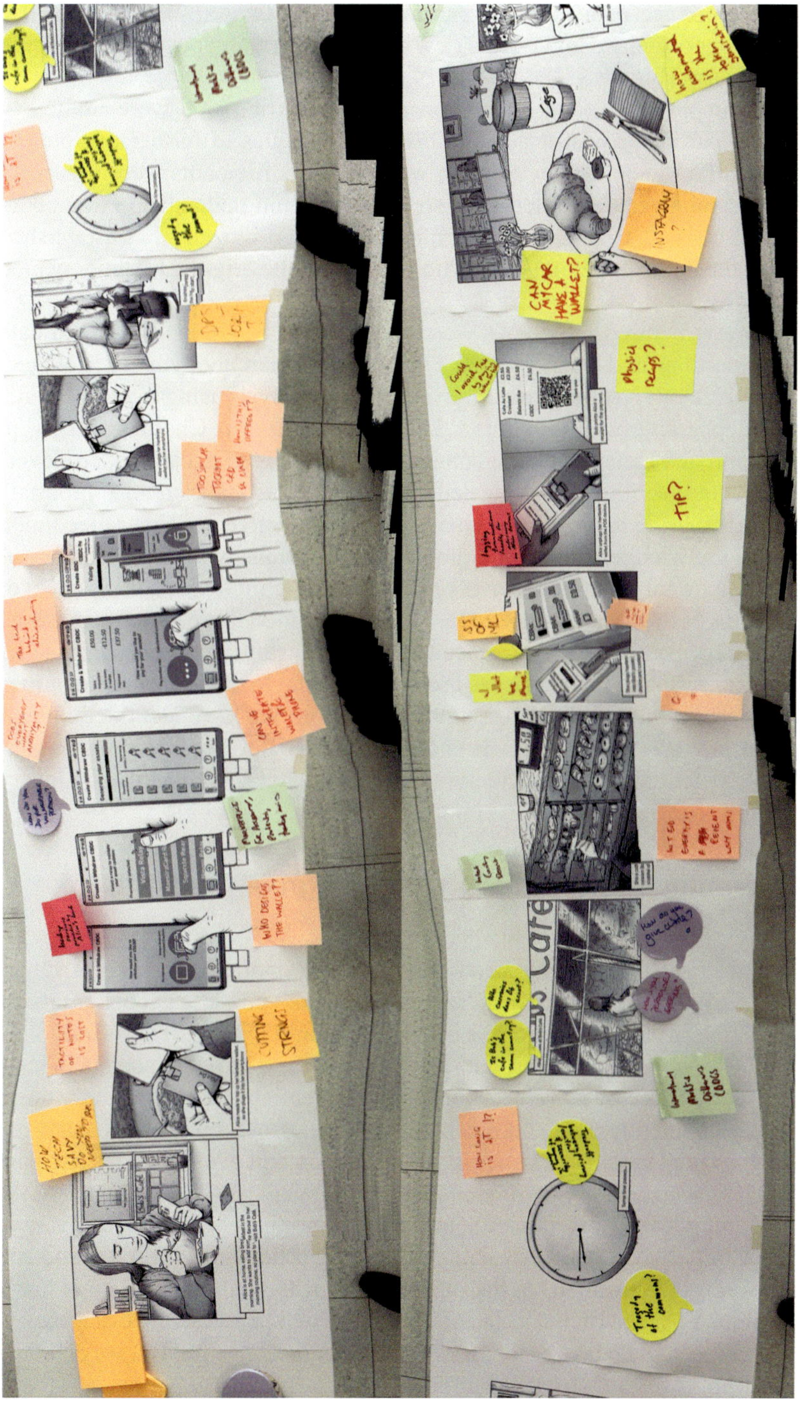

Fig. 5.2 An illustrated comic portrayed a fictional use of CBDC, and participants at two workshops were invited to comment on the strengths, weaknesses, opportunities and threats that the technology may have for citizens. Photograph by the authors.

to Alice's experience using CBDC, participants were invited to post strengths, weaknesses, opportunities and threats (SWOT analysis) using post-it notes on top of the comic strip (Figure 5.3).

During a 'playback' and discussion session, participants were invited to expand upon particular post-it notes to enable colleagues to further explore the implications of an idea. Following both workshops, participants were invited to take part in semi-structured interviews so that we would understand more about the pros and cons of CBDC and its implications upon citizens and established members of the finance sector. Four participants from the original fifteen accepted the opportunity.

The semi-structured interviews were conducted remotely and captured data via audio recording and transcription. The data was then anonymised, collated, and uploaded into Nvivo 14 for qualitative thematic analysis using a constructivist grounded theory approach (Charmaz, 2006). According to Charmaz (2014, p. 131), this approach can reveal "the connections between micro and macro levels of analysis" and link "the subjective and the social." This makes it an appropriate method for extracting how an individual's micro subjective insights reflect upon macro social knowledge, values, and global considerations with the emergence of CBDC. The data analysis process involved conducting a line-by-line assessment of each transcript to formulate initial codes. We then refined these codes through an axial coding process and cross-referenced their relationships with codes from other participant transcripts. Finally, we used selective coding to identify core categories and present emergent theories in our findings section. We have identified themes that can aid interaction designers in understanding CBDC from various angles, including its potential for value creation and inclusivity as well as its cautionary tales. The following section reports on these findings.

4. Reforming Values through the Design of Central Bank Digital Currency

Our research suggests that CBDC presents a unique opportunity to reform values through its design. As one participant stated, "We're not going to get better money with CBDC unless we design it to be a better one" (Participant 3). This is not solely due to CBDC's technical abilities, but rather the chance it provides society to have money represent more of its values and desirable practices than it currently does.

One participant referred to current monetary infrastructures as "haphazard and a bit disorganised," leading to issues like "oppression and exclusion" (Participant 1). CBDC was seen as an opportune moment to reinvent the values of money. As one participant put it, "Let's think about the role that digital currency can play in creating the sort of society that we actually want" (Participant 1). Yet, time to add such values might be limited, as the introduction of CBDC into society is just a matter of time. As one participant stated, "The stone will be thrown. We don't know where it's gonna land and we don't know how deep it's gonna go. But we do know there'll be ripples from it" (Participant 4).

Fig. 5.3 Participants discussed the implications both during and following the workshops. Photograph by the authors.

The unknown ripples of CBDC imply its trajectory and values will only become clearer as it comes into being. Yet, some predictions might be afforded to its positive value impact. For instance, through its grounding in DLT and blockchain technology, CBDCs could redefine trust in the economic realm and track commercial practices from production to consumption using smart technologies and influence behaviour for the better (Powell et al., 2023). As one participant explained, "Digitalisation of trade [. . .] to increase the right sort of growth and enable greater demonstration by companies and traders of their ESG compliance and good ethical and carbon neutral behaviour" (Participant 1). This was seen as an opportunity to offer companies VAT relief for services that are not carbon heavy. Essentially, "You could affect people's behaviour for the better" (Participant 1).

The potential for CBDC to influence behaviour, while potentially positive, also presents challenging futures. As one participant stated, "Some of the folks feel that we have already given away so much of our privacy that it wouldn't be a problem to have a central entity like the Central Bank, [. . .] you know, take a peek on, you know how many pints I bought tonight. [. . .] Do we want to have that at the core of our financial system?" (Participant 3). CBDC can do wonderful things for concepts like sustainability, but it can also penalise people if their practices do not align with the governed values present in the CBDC design, potentially leading to harmful practices. As one participant expressed concern, "I'm concerned about privacy and protections [. . .] we need to prevent abuses from happening" (Participant 2).

For participants, CBDC was not just an aesthetic change to money in digital form but an opportunity to reinvent its social representations that could drive new notions of inclusive, ethical and global morals. Nevertheless, it is evident CBDC requires a fine balance between what and whose values are being re-imagined and equally whose are not. If history gives a sense of future risks and challenges, then vested interests in the economy are certain to use their standing and lobbying efforts to ensure the introduction of CBDCs will be in their commercial interest and favour their private profitability over public good aspirations (Hosseini and Gills, 2024; Hakken et al., 2015; Diesendorf and Taylor, 2023).

5. The Consideration of Time in the Design and Deployment of CBDC

The speed at which a CBDC is deployed, discussed, and regulated can impact the values reflected in its design. "I feel like it's been a very rapid acceleration to where we are. It's been very focused on the kind of the consultation paper reading [from the Bank of England]." (Participant 4). An echoed reflection: "The debate about CBDC has got [too] specific [too] quickly in my view and is currently mouldering in the regulatory space, and it really shouldn't be there at all. This should be a civil society debate" (Participant 1). Rushing the process may limit the involvement of society in the conversation.

Aiming to be first or reacting to the progress of others could also result in a CBDC that does not fully represent the values of its users: "Half-baked because you haven't given it enough thought" (Participant 3). Having time to think through a value-centred approach (Nathan et al., 2008; Carter, 2022) and ensuring that the CBDC represents ethical, sustainable, and future-oriented values could lead to greater uptake of a digital coin that "People will want to use to demonstrate their ethical, their sustainable, and their future" (Participant 1). Pace becomes an important consideration in the design of a CBDC. Taking the time to ensure that values and needs are built into the design is necessary, because if these values and needs are not incorporated, then uptake of the system may be jeopardised.

One current value of money is that moving it around can be slowed and "limited by the movement of physical paper currency," according to Participant 2. This can be positive, for example, it can inhibit a run on another central bank's currency. Participant 2 explained that central banks often hold each other's currencies. However, if this action becomes more CBDC-like than physical asset-based, then moving money becomes instantaneous, posing a problem. "So you then have the situation where a central bank is no longer able to prevent a run on its currency because other central banks can just send them off to other partners at will without another central bank being able to stop it," said Participant 2.

The quickness of a CBDC was also seen as a positive. "Frictionless experience for somebody who's wanting to trade across a border, especially for the small to medium enterprise sector, who so far are showing little appetite for doing it," said Participant 1. While making a CBDC's speed of transaction and the associated frictions—or lack thereof—a matter of context, previous research suggests there is merit in the explicit use of frictions to encourage mindful interactions (Cox et al., 2016).

6. Echoing Present Values in the Design of Central Bank Digital Currency

The present values associated with money are vast and diverse. They carry specific personal or social values with them in addition to the commercial values by institutional and corporate partners, represented by our study participants. This suggests that questions such as the one asked by Participant 4 become important in CBDC discourse and design considerations: "How do we get the value of the automation from digital currency without losing the community sense that you get from cash, you know, how it interacts with the community?" Cash became a signpost to echo present values. "Immediately and anonymously, I can get cash out of the bank [...] I can give somebody a £10 note and my debt is immediately anonymously and frictionlessly settled" (Participant 1). Cash offers privacy, immediacy and other values. However, a CBDC does not necessitate the removal of cash altogether. "I think there's going to be less need for cash, but there's still going to be a need for it" (Participant 2).

Cash offers other "quality differences" (Participant 2) that ought to be considered in the design of CBDCs. "Cash for example is more efficient to budget on. So it's more important to have that during times of financial crises.

So I'm hoping [...] a digital pound would serve those kinds of financial budgeting capabilities and enable people to better control their money" (Participant 4). Cash has many values connected to it that are desirable and beneficial for a CBDC to try to replicate.

Generationally people have come to form an idea of their money. "You know, paper money because there are still generations that are used to that. There is a cultural, societal aspect to it that values that right and the shocker to me when I arrived in the UK was like places that only accepted contactless payments. In Brazil, if you're a business and you don't accept cash, bank notes, it's a federal crime" (Participant 3). People's present values for cash might rely on explicit protection from policy and law, as suggested by the Reserve Bank of New Zealand (2021).

The social values embodied in the current use of cash might slow down its exit from society: "It would be a real shame if the only outcome of CBDC was as a wholesale one, and it made [Commercial bank] 1 per cent more profit. That would be a real shame" (Participant 1). The present gives an idea for the things a CBDC might want to implement in its design as well as indicate what to avoid or try to change for the better.

7. Promoting Inclusion in the Deployment of Central Bank Digital Currency

The type of CBDC that emerges is important in determining who and how we consider the inclusionary properties afforded or not afforded by CBDC. Some participants delineated between two communities to be considered in this process, firstly 'wholesale' which refers to banking services sold to large clients, and secondly 'retail' users who represent the large populace. Participant 3 explains that "If it's wholesale [CBDC], you're talking to people that are in the top 0.1 per cent of capitalism. [. . .] I don't think we need to involve, you know, John Smith from down the street to discuss wholesale CBDC." Wholesale and retail CBDC were seen to require different user inputs. Moreover, the extent to which inclusion was practiced was dependent on its deployment, whether done in the commercial bank sector or centrally distributed.

"Imagine you're a commercial bank, right? You have your demographic that, you know [. . .] You have a smaller population than the whole central bank would have [. . .] You know your demographics that you can tap into at your scale and you can design for them a tokenised deposit experience that makes sense to those customers" (Participant 3). On the commercial side of banking, a user focus was seen to be an agreed consensus. "I think customer centricity is vital. You know as early as possible" (Participant 4). However, Participant 4 stated "I think the user focus is: 'No, we don't [focus on users]'" (Participant 4). This was because, "We are very early on, knowing around [about] what the CBDC does" (Participant 4). Grasping an idea of what the CBDC does and what the government, market, and economy decide to do with CBDC directly influences which users become central to its inclusive potential. Equally important is considering which users will get left behind with the ways CBDC is deployed and updated. Being digital, CBDC will require updates, security patches, and so on. User focus becomes important here as pointed out by Participant 2: "How do we update it [CBDC] [. . .] Do you get people to convert from that version to another version? Do you fork it into a version that uses a new algorithm? Then what happens to the users that get left behind or don't know about it?" It is clear that the type of CBDC used, its set of versions, and the mechanics of its deployment all become avenues to question whether its design is appropriately user-focused along with the question of what user focus is required and where within the wider notions of CBDC discourse. With regards to how CBDC is discussed, the language being jargon-free aligned with an inclusionary consideration by Participant 4: "I think what is really important for the progress of this, is the inclusiveness of the definitions of the language. I think that is crucial. That is vital. That you need to have everyone agree that we use the same language and it's inclusive. It's not jargony." Ensuring inclusive access to both discussion, education and user-friendliness will require effort in all areas, from the design of CBDC user interfaces to the language, to education approaches used in and around CBDC discourse.

Conclusions

This chapter aims to start a conversation about the inclusivity, value, and potential of CBDC from multiple perspectives. However, this study has some limitations. One is the small sample size of only four participants. Further research with a larger and more diverse group of participants, including those with and without prior knowledge of CBDC, could provide more insights. Another limitation is that all participants had some familiarity with CBDC. Including individuals without prior knowledge of CBDC in future research could broaden our understanding of how it is perceived and reveal new concepts of value and the materiality of money.

This chapter also aims to initiate a discussion among designers about their role in influencing or preserving the values of money, both in its traditional forms and in its emerging form as Central Bank Digital Currencies. Our findings, though limited in scope, reveal the challenges and opportunities that CBDC poses for design. They underscore the potential of design to shape this evolving landscape. We have addressed certain areas where future work could be undertaken by designers. However, in the following we break this down more precisely, addressing designers in the process.

CBDC presents an opportunity to shape values. However, this must be approached with caution. Our findings suggest that people may have differing ideas about what constitutes value. If one group has control over creating or upholding value, others may be subject to oppressive experiences. Design methodologies such as participatory and co-design should be utilised to bring partners within a financial ecosystem together and facilitate discussions between a diverse range of people, including those driving CBDC developments and those who have yet to encounter the concept.

Our study found that bringing together a diverse range of people to discuss CBDC may present challenges due to the topic's complexity. As Participant 4 pointed out, it is important to start building an inclusive language around CBDC. Design may play a role in shaping and helping to create the lexicon associated with CBDC. This could involve designing materials, toolkits, or prototypes that enable people to build a language for discussing CBDC or create a CBDC language that is understandable to many.

The uncertainty surrounding the nature of CBDC, whether it will be wholesale, retail, or based on deposits makes designing for CBDC a challenging task. Yet, as previous designers and researchers have shown (Pschetz and Bastian, 2018; Akama et al., 2015; Sousa at al., 2021) uncertainty can be a useful phenomenon to have within the design process. Design methods, such as speculative design or design fiction could be utilised to envision uncertain scenarios and futures where each of these versions comes to light (Murray-Rust et al., 2023; Blythe, 2014; Rüller et al., 2022; Dunne and Raby, 2013; Galloway and Caudwell, 2018).

Maintaining existing values and established practices associated with money will play a crucial role in shaping the design of CBDC. Factors such as trust, policy, familiarity, and experience will be key considerations for CBDC design. This will be a fine balance for CBDC, as removing existing values could be detrimental to the more than numeric value people place within the discourses of money maintaining their monetary routines (Vines et al., 2012) and having the infrastructure to access their tangible assets (Trotter et al., 2020).

Future work might involve designers finding ways to better understand existing values around monetary discourse and ensuring that CBDC reflects those values through its design.

Designers may want to consider Temporal Design (Pschetz and Bastian, 2018) as a way to explore the various temporal properties associated with CBDC. For example, Participant 1 mentioned that CBDC has too quickly become present in certain spaces. CBDC has varying speeds of operation and research capabilities associated with it. CBDC not being a tangible material means its temporal qualities are very different to that of cash. It has the potential to transfer sovereign money at unprecedented speeds, requiring friction as an important design consideration of CBDC.

Currently, the discourse around CBDC tends to lack a focus on inclusivity. Although some efforts have been made to understand inclusion and exclusion to a great extent, these have largely been theoretical and have not incorporated user-driven perspectives (Barzyk et al., 2022). A theme that emerged from our findings is the connection between value and inclusion and the need for a more user-centric CBDC design. There are many challenges to overcome in this regard, including the speed at which topics are evolving faster than research can keep up (Auer et al., 2021), the diversification of values around money, the use of jargon language, education, and issues identified in the literature such as digital exclusion (Lloyds Banking Group, 2022; Dai et al., 2023; Gopane, 2019), which correlates with concepts of financial inclusion. CBDC is more than just a new form of money—its potential could be significantly positive, as reported by Participant 1. However, without user-led design, its potential to be detrimental is equally significant.

Money has taken many forms throughout history, including non-tangible representations. The rise of Central Bank Digital Currency (CBDC) will transform our understanding of digital currency. Money is connected to many values, numeric, social, cultural, religious, moral and more. However, new forms of money can lead to the demise of other notions and their associated values. This is an unprecedented moment to re-envision the meaning of money. We may need time to consider how money came to be and begin "reverse engineering" (Participant 3) before we can understand what values CBDC does and does not represent and more importantly, whose values are represented by CBDC. Design has an opportune moment right now to shape the ripples made by early CBDC developments and pilots and help establish a more equitable and inclusive social, political and global foundation associated with money. It is important to consider who has a stake in this most common design material, money, in its newly emerging form, digital sovereign currency.

Acknowledgements

The authors acknowledge funding from the UKRI / EPSRC research councils and the PETRAS Research Centre EP/S035362/1 for the FIRE Project. The authors would also like to thank all the participants for their engagement with the research.

References

Akama, Y., Pink, S. and Fergusson, A. (2015). Design + Ethnography + Futures: Surrendering in Uncertainty. *Proceedings of the 33rd Annual ACM Conference Extended Abstracts on Human Factors in Computing Systems (CHI EA '15) Association for Computing Machinery, New York, NY, USA*, 531–42. https://doi.org/10.1145/2702613.2732499.

Arvidsson, N. (2018). The future of cash. In *The Rise and Development of FinTech* (pp. 85–98). Routledge. https://doi.org/10.4324/9781351183628.

Arvidsson, N. (2019). *Building a Cashless Society: The Swedish Route to the Future of Cash Payments*. Springer Briefs in Economics. Springer International Publishing. https://doi.org/10.1007/978-3-030-10689-8.

Auer, R., Frost, J., Gambacorta, L., Monnet, C., Rice, T. and Shin, H. S. (2021). *Central bank digital currencies: motives, economic implications and the research frontier* (Technical Report 976). Bank for International Settlements. https://doi.org/10.1146/annurev-economics-051420-020324.

Barzyk, K.M., Macek, V. and Giraldo Mora, J.C. (2022) Inclusive and exclusive affordances of central bank digital currencies: A multiple-case study of Sweden and Bahamas. https://research.cbs.dk/files/76450614/1397869_Master_Thesis_Karol_Viktor_Main_Doc.pdf.

Baumer, E.P.S., Ames, M.G., Brubaker, J.R., Burrell, J. and Dourish, P. (2014). Refusing, limiting, departing: why we should study technology non-use. *CHI '14 Extended Abstracts on Human Factors in Computing Systems*, 65–68. https://doi.org/10.1145/2559206.2559224.

Blythe, M. (2014). Research through design fiction: narrative in real and imaginary abstracts. *Proceedings of the SIGCHI Conference on Human Factors in Computing Systems*, 703–712. https://doi.org/10.1145/2556288.2557098.

Ceeney, N. (2019). Access to cash review. Final Report, March 2019. https://www.accesstocash.org.uk/media/1087/final-report-final-web.pdf.

Centellegher, S., Miritello, G., Villatoro, D., Parameshwar, D., Lepri, B. and Oliver, N. (2018). Mobile Money: Understanding and Predicting its Adoption and Use in a Developing Economy. Proc. ACM Interact. Mob. Wearable Ubiquitous Technol. 2, 4, Article 157 (December 2018), 18 pages. https://doi.org/10.1145/3287035.

Charmaz, K. (2006). *Constructing Grounded Theory: A Practical Guide through Qualitative Analysis*. Sage.

Dai, J., Miedema, J., Hernandez, S., Sutton-Lalani, A. and Moffatt, K. (2023). Cognitive Accessibility of Digital Payments: A Literature Review. *Proceedings of the 20th International Web for All Conference (W4A '23), New York, NY, USA, 116–121*. Association for Computing Machinery. https://doi.org/10.1145/3587281.3587294.

Desan, C. (2021). Money's Design Elements: Debt, Liquidity, and the Pledge of Value from Medieval Coin to Modern "Repo" (August 1, 2021). Banking and Finance Law Review, v. 38 (2022 Forthcoming), Harvard Public Law Working Paper No. 21-31, Available at SSRN: https://ssrn.com/abstract=3897399.

Dunne, A. and Raby, F. (2013). *Speculative Everything: Design, Fiction, and Social Dreaming*. MIT Press.

Galloway, A. and Caudwell, C. (2018). Speculative design as research method. In G. Coombs, A. McNamara, & G. Sade (Eds.), *Undesign: Critical Practice at the Intersection of Art and Design* (1st ed., pp. 85–96). Routledge. https://doi.org/10.4324/9781315526379-8.

Gopane, T.J. (2019). An Enquiry into Digital Inequality Implications for Central Bank Digital Currency. *Proceedings of 2019 IST-Africa Week Conference (IST-Africa), Nairobi, Kenya, 2019*, 1–9. http://dx.doi.org/10.24193/subbnegotia.2024.1.01.

Halloluwa, T., Bandara, P., Usoof, H. and Vyas, D. (2018). Value for money: co-designing with underbanked women from rural Sri Lanka. *Proceedings of the 30th Australian Conference on Computer-Human Interaction (OzCHI '18), New York, NY, USA, 63–73,* 63–73. Association for Computing Machinery. https://doi.org/10.1145/3292147.3292157.

HM Treasury and Bank of England. (2023). The digital pound: A new form of money for households and businesses? https://www.gov.uk/government/consultations/the-digital-pound-a-new-form-of-money-for-households-and-businesses.

Ingham, G. (1996). Money is a social relation. *Review of Social Economy*, *54*(4), 507–29.

Ingham, G. (2013). *The Nature of Money*. John Wiley & Sons.

Iqbal, M.Z. and Campbell, A. (2020). The emerging need for touchless interaction technologies. *Interactions*, *27*(4), 51–52. https://doi.org/10.1145/3406100.

Lloyds Banking Group. (2022). 2022 consumer digital index. https://www.lloydsbank.com/assets/media/pdfs/bankingwithus/whats-happening/LB-Consumer- Digital-Index-2022.pdf.

Maurer, B. (2006). The Anthropology of Money. *Annual Review of Anthropology*, *35*, 15–36. https://doi.org/10.1146/annurev.anthro.35.081705.123127.

Murray-Rust, D., Elsden, C., Nissen, B., Tallyn, E., Pschetz, L. and Speed, C. (2023) Blockchain and Beyond: Understanding Blockchains Through Prototypes and Public Engagement. ACM Trans. Comput.-Hum. Interact. 29, 5, Article 41 (October 2022), 73 pages. https://doi.org/10.1145/3503462.

Peneder, M. (2022) Digitization and the evolution of money as a social technology of account. J Evol Econ 32, 175–203 (2022). https://doi.org/10.1007/s00191-021-00729-4.

Powell, W., Cao, S., Foth, M., He, S., Turner-Morris, C. and Li, M. (2023). Revisiting trust in supply chains: How does blockchain redefine trust? In B. Abdelaziz, I. Khalil, & B. Aouni (Eds.), *Blockchain Driven Supply Chains and Enterprise Information Systems* (pp. 21–42). Springer Nature. 10.1007/978-3-030-96154-1_2.

Pschetz, L. and Bastian, M. (2018) Temporal design: Rethinking time in design. *Design Studies*, *56*, 169–84. https://doi.org/10.1016/j.destud.2017.10.007.

Reserve Bank of New Zealand (2021). The Future of Money – Stewardship. Public consultation paper, September 2021. https://web.archive.org/web/20221106083942/https://www.rbnz.govt.nz/-/media/project/sites/rbnz/files/consultations/banks/future-of-money/stewardship-issues-paper.pdf.

Roitman, J. (2023). Platform economies: Beyond the North-South divide. *Finance and Society*, *9*(1), 1–13.10.2218/finsoc.8089.

Rüller, S., Giménez Ciciolli, B., Weibert, A., Aal, K., Blythe, M., Muller, M. and Kafai, Y.B. (2022). Creative Entrances to Co-Design: Exploring Collaboration through Fiction, Fairy tales, and Games. *Proceedings of the Participatory Design Conference 2022 - Volume 2*, 247–50. https://doi.org/10.1145/3537797.3537877.

Satchell, C. and Dourish, P. (2009). Beyond the user: use and non-use in HCI. In *Proceedings of the 21st Annual Conference of the Australian Computer-Human Interaction Special Interest Group: Design: Open 24/7 (OZCHI '09) New York, NY, USA*, 9–16. Association for Computing Machinery. https://doi.org/10.1145/1738826.1738829.

Schindler, J. (2017) *Fintech and financial innovation: Drivers and depth* (Technical report). Board of Governors of the Federal Reserve System (US). https://doi.org/10.17016/FEDS.2017.081.

Selwyn, N. (2003). Apart from technology: understanding people's non-use of information and communication technologies in everyday life. *Technology in Society*, *25*(1), 99–116. https://doi.org/10.1016/S0160-791X(02)00062-3.

Sousa, S.P.B., Baptista, A.J. and Marques, A.T. (2021). Material Design-for-Excellence Framework – Application to Composites. 10.1016/B978-0-12-819724-0.00105-1.

Speed, C., Rankin, J., Elsden, C. and Vines, J. (2022). The future of money as a design material. In Lockton, D., Lenzi, S., Hekkert, P., Oak, A., Sádaba, J., Lloyd, P. (eds.), DRS2022: Bilbao, 25 June - 3 July, Bilbao, Spain. https://doi.org/10.21606/drs.2022.785.

Swierstra, T. (2013). Nanotechnology and technomoral change. *Etica & Politica/Ethics & Politics*, *XV*(1), 200–219.

Wiberg, M. (2020). On physical and social distancing: reflections on moving just about everything online amid COVID-19. *Interactions*, *27*(4), 38–41. https://doi.org/10.1145/3404213.

6 Blockchain and Tokenisation for Local Communities

Guido Boella[1*], Claudio Schifanella[2] and Cristina Viano[3]

1. Introduction

The main innovation introduced by blockchain is the possibility of creating cryptographic tokens (digital representations of a value or property right over an asset) and its ability to enable safe and transparent transactions of these tokens without relying on third parties as intermediaries. Moreover, smart contracts (software that operate over a blockchain, automatically executing the terms of an agreement when certain conditions are met) allow for programming the way in which tokens are transferred, thus automatising transactions. Hence, the potential of blockchains to disintermediate governance and business models. Recent years have seen the rise of experimentation with *blockchains for social impact* or *blockchains for social good,* i.e. blockchain-enabled activities that tackle urgent societal and environmental needs (Galen et al., 2018, 2019; Polvora et al., 2020; Voshmgir et al., 2019). The report "Blockchain and Sustainable Development" (Fines Schlumberger, 2022) mentions 14 categories of blockchain applications[#] ranging from productive sectors to health, education and public administrative domains. Less experimentation is taking place in a domain that cross-cuts three of the above categories, namely Peer-to-Peer Cash; Government & Democracy; and Production and Consumption. This overarching domain is about the participation of citizens in the civic, democratic and economic life of their local community. Processes of co-production of public services by public administrations and citizens (Brandsen and Honingh, 2018), as well as citizen self-organisation initiatives such as urban commons (Bollier & Helfrich, 2019; Foster & Iaione, 2016) and local solidarity economies (Miller, 2010; Vlachokyriakos et al., 2017), are crucial for fairer and just societies, and blockchains can support and enhance such existing participatory processes or enable new ones. Cities are organised in interconnected communities

[1,2,3] Dept of Computer Science, University of Turin, C.so Svizzera 185-10149, Turin, Italy.
Email: guido.boella@unito.it[1]; claudio.schifanella@unito.it[2]; cristina.viano@unito.it[3]

* Corresponding author: guido.boella@unito.it

[#] Identity & Property; Peer-to-Peer Electronic Cash System & Programmable Money; Health; Supply Chain & Logistics; Energy; Aid, Charity & Philanthropy; Agriculture & Food; Web 3.0, Art & Sciences; Environment & Climate; Government & Democracy; Electronic Communication & Telecommunications; Education & Employment; Production and Consumption; Infrastructure and Transports.

composed of citizens, economic activities and associations, each governed and managed with different rules and tools but in continuous interaction with the others. Tokenisation (the conversion of the right to an asset into a cryptographic token)[1], smart contracts and tokenised economies (that leverage tokenization mechanisms for circulating values and incentives) can provide common and interoperable tools as a relevant layer to represent community rules and create new intercommunity services.

While there are various blockchains to support decentralised management of physical and digital infrastructures in (smart) cities (Bagloee et al., 2021; Shen & Pena-Mora, 2018; Xie et al., 2019), tokenisation is less significantly exploited to facilitate civic participation and social economies. However, researchers in the HCI and design fields are considering blockchains in civic contexts, and they are getting insights on how citizens perceive their potentialities and risks (Nissen et al., 2018; Tallyn et al., 2020; Cila et al., 2020; Murray-Rust et al., 2022). Moreover, recent projects experiment with blockchain for local communities, mainly in the form of systems for rewarding civic participation and fostering local economic activities, or of community currencies and tools for financial inclusions: for instance the projects Cambiatus, Colu, UrbanChange, Circles, Sarafu Network, ImpactMarket, GoodDollar (see Domenicale et al., 2024 for a review on their purposes and tokenized economy models). These projects differ from applications for decentralised finance[2], and from speculative cryptocurrency exchanges in terms of incentive mechanisms, governance models, and scale of implementation. They represent a new direction towards blockchain designed for a positive societal impact as they aim to implement financial schemes that offer alternatives to speculative cryptocurrencies in global financial markets. However, all these projects revolve around a specific model of a tokenised economy, codified in an application, with few opportunities for interoperability with similar tools and adaptability to different local socio-economic contexts and needs. There is a lack of effective methodologies and tools for the design of tokenised economies according to the specific needs, not only of social economies and civic participatory processes (Vlachokyriakos et al, 2016), but also of different local communities. More evidence is needed regarding the potentials and challenges of customizable blockchain-based tools for the pursuit of just and fair local communities.

---

1 The term tokenisation should not be confused with tokenism as used in domains such as diversity and inclusion, or urban planning. Tokenism refers to actions that should provide people or groups of citizens with opportunities for substantial participation, but in practice do not give them real power (Monno and Khakee, 2012). This can happen, for instance with the inclusion of specific categories of workers and with the political representation of minorities, but also with participatory research and the co-production of public services when citizens are merely consulted and do not actually influence decision making.

2 Decentralized finance (DeFi) is a set of financial instruments that leverage smart contracts and cryptocurrencies, thus avoiding intermediaries such as banks or brokerages.

Most of the local contexts where civic participation and social economies take place in a digitally-enabled way are urban. Indeed, cities are a privileged domain for digital social innovation (Rodrigo et al., 2019) due to their density of socio-economic relationships (Certomà, 2021). Hence, our focus is on *urban* local communities as places where blockchains can usefully support and enable participatory and inclusive processes. Many local governments are making an effort, not only in e-government, i.e. digitising services and optimising work processes, but also in digitally-enabled co-production of services of public interest (Lember, 2018; Yuan, 2019). Co-design and co-production reimagines the role that citizens have in public service creation, production, and management in collaboration with public officers, and is expected to make policies and services more responsive to actual individual and social needs (Estrada-Grajales et al., 2018; Lewis et al., 2020). Meanwhile, civil society organisations, community members and digital activists are experimenting with a variety of grassroots urban projects that leverage digital technologies for the purpose of solidarity practices, social and collaborative economies, and urban commoning, and they are motivated by visions that contest the mainstream approach to smart cities (Vadiati, 2022). As Olivier and Wright (2015, p. 62) put it: "In these times of austerity, there is a risk that digital civics might be construed, on the one hand, as finding ways of making citizens do it for themselves, or on the other hand, as dismantling public service provision." Our approach to digitalization goes beyond government cost-saving and making processes more efficient. Rather, it aims at enabling co-production practices where interactions between government and citizens are relational as opposed to (only) transactional (Boyte, 2004; Olivier & Wright, 2015; Vlachokyriakos, 2017).

This chapter is based on a research project that investigated the potential of blockchain and other "disruptive" technologies to facilitate new ways of managing co-production of public services and civic participation in urban communities. The project "CO3 - Digital Disruptive Technologies to Co-create, Co-produce and Co-manage Open Public Services along with Citizens" (hereinafter CO3) addressed different digital technologies in different phases of co-design, co-production and co-management of public services. Four technologies (blockchain, geolocated social networks, interactive democracy[3] and augmented reality) and gamification techniques were integrated in a single mobile app in order to go beyond information exchange enabled by social networks and provide new forms of interaction. Geolocated civic social networks enhance the local and civic dimensions of the exchange of information and coordination. Gamification is a fun way to encourage and acknowledge participation and civic engagement (Rehm et al., 2018). Interactive democracy gives participants transparency and a voice. Augmented reality (AR) provides an immersive interface for accessing CO3 functionalities. The blockchain plays a crucial role in that it is leveraged not only for transactions of money but also goods and services, thus adding the dimension

[3] Interactive democracy refers to the use of the internet to facilitate new forms of democratic participation, from making proposals to discussion, option formation, voting and deliberation.

of exchange of different types of values. The relationship between the mobile app and the urban context is grounded on the conceptual framework of an Augmented Commoning Area (ACA). The readiness level of these technologies, the growing interest in the co-production of public services, and the strengthening of urban commons as a reliable social structure in many cities makes such a conceptual framework plausible. The CO3 project was conceived to explore the viability of these technologies for urban commoning, particularly blockchain for economic urban commoning. The pilots activated within the CO3 project were focused on co-design, technical development and testing activities. These activities allowed the participatory assessment of the relevance, viability, strengths and challenges of blockchain-based ACA, considering both socio-cultural and technical factors.

The remainder of this chapter is structured as follows: The notion of an ACA is introduced in Section 2, followed by insights on the different types of interactions enabled by blockchain and tokenised economies, AR, geolocation, interactive democracy, social networking and gamification. In Section 3 we describe how the ACA concept has been contextualized in different scenarios through the pilots of the CO3 project. In Sections 4 and 5 we discuss the potentials and challenges of ACAs and tokenized commoning, and we present further developments of blockchain-based wallets for local communities, based on the findings of the CO3 project.

2. Augmented Urban Commoning

2.1 Augmented Commoning Areas

The CO3 project refers to commoning as a set of processes through which different actors cooperate for creating and managing services of public interest, social activities, shared spaces and urban commons. In particular we refer to cases in which local public administrations are rediscovering the pre-modern practices of the commons (Ostrom, 1990) as an alternative model of co-production where solidarity, inclusion and public-private cooperation are crucial. Urban commons represent a concrete framework-solution to give municipalities the potential to create new forms of welfare, exploiting practices of co-creation and co-management of services such as cultural hubs, urban garden, educational services, self-managed kindergartens, and assistance to the elderly[4].

The core concept of the CO3 project is augmented commoning. An augmented commoning area (ACA) is a 'phygital' (physical and digital) public space where urban actors can engage in co-production processes through digitally-enabled interactions (see Figure 6.1). Augmented Commoning Areas (ACAs) are defined by geographical boundaries in physical urban spaces. The enabling technological

[4] One example is the Italian Public Regulation on Collaboration for the Care and Regeneration of Urban Commons that enables agreements (the Pacts of Collaboration) to be made between citizens and public authorities for the creation of services in urban commons. The city of Turin is among the Italian cities that adopted such a regulation.

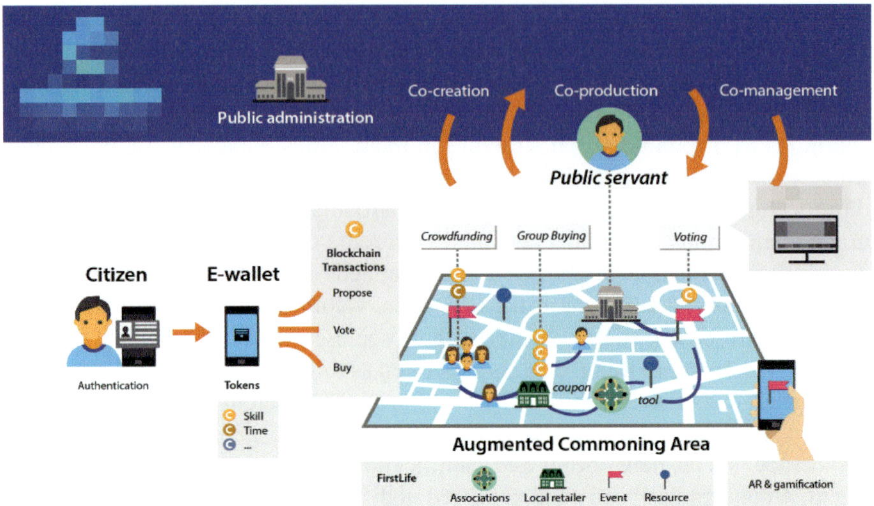

Fig. 6.1 Visual representation of the ACA concept. Image created by authors.

functionalities are provided by the CO3 technological platform and are accessible when the participant is physically present. Different technologies support the interchange of intangible resources (creativity, information, knowledge, wants, needs, opinions, feedback) and tangible resources (objects, work, time, values) through their digital representation as cryptographic tokens. In the following sections, we will explain how CO3 tokenised economic interactions enabled by blockchain are oriented towards co-production; how they are linked to urban spaces with augmented reality and geolocation; and how they are integrated with other social interactions with interactive democracy, social networking and gamification.

The ACA concept distances itself from common public service web platforms that often provide a less than engaging user experience. CO3 augmented commoning areas (ACAs) are more ambitious and have the following aims:

- augmenting opportunities to share and exchange information to stimulate cooperation between commoners and with public administrations;
- enabling new practices of solidarity, volunteering and commoning through new digital economic mechanisms;
- improving people's wellbeing by mobilising existing resources both tangible (e.g. fresh food) and intangible (time, skills);
- enabling systematic mechanisms for cooperative decision-making related to urban commons.

The five disruptive technologies presented below have been integrated in the CO3 platform to facilitate interaction between participants in the context of specific scenarios implemented at three pilot sites. As the five CO3 applications use diverse conceptual models and lexical formats to represent the data they manage, these have to be reconciled to support cross-application data sharing. The core interoperability between the different components of the CO3 platform is provided by FirstLife, used as a shared interchange communication layer to allow cross-application logging of user actions and tracking of user interaction with the various services offered. Moreover, the FirstLife ontology is used as an interlingua and defines a unified information representation format for geographic and non-geographic data. With FirstLife, there is no need to have different communication application programming interfaces (APIs) for each pair of applications. And it allows us to avoid common concepts being represented differently in different applications. The primary implementation of the CO3 platform is as a mobile application that runs on both Android and iOS devices. This is complemented by a blockchain wallet, which is implemented as a progressive web application.

2.2 Blockchain-based Social Economies

One of the opportunities provided by the integration of blockchain within the CO3 project is its use as a notarisation tool. Due to its time-ordered structure, strong resistance against tampering attacks, and transparency, blockchain is an ideal notarisation tool to record information in a secure and verifiable manner. Moreover, the use of smart contracts (self- enforcing agreements written in code) to create and distribute tokens that implement highly customisable and automated value exchanges is an attractive mechanism that can create trust among users, and encourage them to establish new forms of collaborative work. As regards more specifically co-creation and innovation processes, an interesting application is the interchange of intangible goods, such as ideas and property rights, by tokenising them, thus making them liquid and transferable.

In the CO3 project, participants access and use the blockchain with a personal wallet. The enabler of transactions in CO3 are fungible tokens as coins and non-fungible tokens as coupons, both represented as standard ERC20-compliant contracts. *Coins* model the behaviour of complementary currencies. *Coupons* are used as digital twins of single objects like goods, discount coupons or access to events. The creator of the tokens and coupons decide their values and, in the case of coupons, what they represent or offer or give access to. Users either purchase tokens with fiat currency, or they obtain them as gifts or rewards. The *crowdsale* functionality allows local actors to create fundraising or group-buying initiatives where the creator offers coupons that represent access to a good or service or event. Participants contribute by purchasing coupons in advance with coins, and so, in order to participate, they must own a certain amount of the local coins. When the quantity of coins collected reaches a minimum threshold, the crowdsale is successful and the underlying smart contract automatically transfers the coupon(s) from the participants' wallets. The *coupon box* functionality allows users to create virtual boxes containing coupons to be offered to other users, without the need to purchase them with tokens. The coupon box creator decides the number of coupons available without setting a price, nor a goal (contrary to crowdsales). The permission to create new coins, coupons, crowdsales and coupon boxes is restricted to certain users with coordination or other specific roles in the neighbourhood community or in the urban commons. All users can transfer tokens, participate in crowdsales and use coupon boxes. Tokens can be transferred to other users either at the same location by exchanging quick response (QR) codes or remotely by sharing QR deep links or user profiles. Users can view their transaction history on their wallet at any time. Users can access crowdsales and coupon boxes by tapping on their visual representation as AR objects or as entities on the geolocated social network FirstLife map (Section 2.3) integrated within the ACA. Combining these tools and functionalities based on cryptographic tokens and smart contracts, different kinds of tokenised social economies and tokenised commoning processes are enabled, as exemplified in the different scenarios of the pilots in Turin and Athens described in Section 3.

2.3 Augmented Civic Interactions

Transactions of values via the blockchain are complemented by other kinds of social interactions that characterise co-production and commoning processes: information sharing, coordination on joint initiatives, deliberation processes, civic engagement and fostering of civic participation. Such interactions are enabled by specific components of the CO3 platform.

FirstLife is a geolocated civic social network (Boella et al., 2019) that provides an interactive map of the city with information and coordination functionalities based on geographic location. It allows interactions with local dynamics, allowing citizens to connect and interact by proximity, and geolocating contents to specific urban places (streets, buildings, and parks). The main FirstLife interface is composed of a map where participants to an ACA can post "entities," and a side wall displaying summary cards of the different entities. Single entities can be opened by clicking on map markers or on their summary card on the wall: this will open a detailed view that shows all their properties (such as categorisation, description, linked URLs, etc.). Entities are the virtual spaces where users can share proposals, events, news, add second layer resources like comments, attachments, photos etc., and react to contents posted by other users. FirstLife is intended to be used as a coordination instrument by different stakeholders like public administrators, association managers and citizens for sharing information and collaborating on activities that have public relevance at a local and city level. In CO3, for instance, commoners can create a "task" entity to announce activities to be carried out; other community members can take responsibility for the task with a check-in (and check-out) action on the digital entity. Local retailers can create a "coupon box" entity at a specific location which other users can use to redeem the coupon it contains, thanks to its integration with the blockchain wallet component. Users can create other entities like images, news, and crowdsales. FirstLife is a web application accessible with a browser from desktop or mobile devices, and it is integrated as a web view inside the CO3 native mobile app.

Interactive democracy functionalities are provided by LiquidFeedback (Behrens et al., 2014), a mature open-source software that has been under active

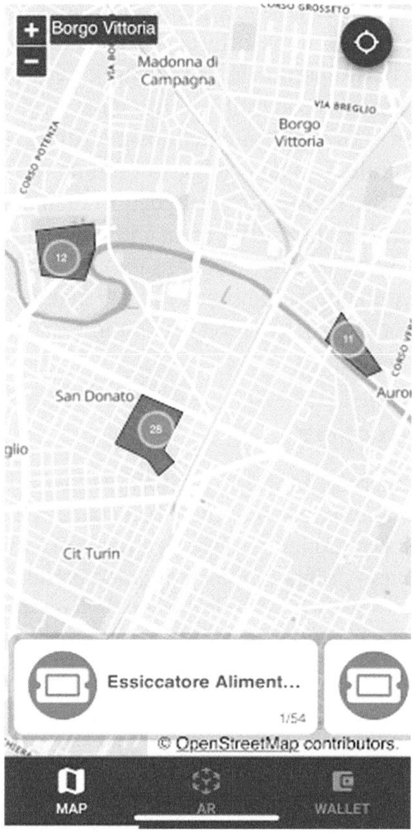

Fig. 6.2 Visualization of ACAs and related entities on the FirstLife map – image created by the authors.

development since 2009. LiquidFeedback enables new forms of democratic participation covering the whole deliberation process, from the introduction of the first draft of a proposal to the final decision. More specifically, the platform enables forms of liquid democracy, giving users the option to directly take part in decision making processes or to choose a representative. During the deliberation, several alternative, sometimes competing, proposals can come up, which are voted upon using preferential voting mechanisms. LiquidFeedback aims to give every participant equal rights and opportunities in the decision-making process and does not allow participants to have any special privileges for moderating a discussion or steering a decision-making process. Participants in the ACA can thus take part in democratic decisions on proposals and activities to be implemented in the urban commons, or on services to be activated within the urban area.

Gamification methods are employed to encourage desired behaviours and positive change with interesting experiences and positive reinforcements (HeLinehanrvas et al., 2015; Rehm et al. 2018). The basic components of a gamified experience, such as points, badges, rewards and progression levels promote and reinforce different types of actions offered by the CO3 platform – like adding content, proposing or participating in public initiatives and deliberations, and exchanging economic assets through the blockchain wallet – and to increase the frequency and persistence of these interactions over time. Progression levels are based on simple points collection and badges. Points are assigned for participation in relevant events. Specific badges are designed to recognise different skills and types of contribution to life and services in the ACA (e.g. augmented content, participation in events, volunteering, wallet usage). There is a particular emphasis on rewarding and incentivising collaborative rather than competitive interactions, and on increasing the level of participation in commoning processes, as exemplified in the Turin pilot (see Section 3).

2.4 Augmented Discovery

The core distinctive feature of the CO3 platform is that it brings together virtual and physical interactions in relation to an urban space. This is made possible by the AR and geolocation functionalities (see Section 2.3), which represent different and complementary access points to the augmented commoning area.

Augmented reality (AR) is a technology that augments our senses with layers of information by combining the physical world with virtual items such as text, images, audio, videos, or objects (Caudell & Mizell, 1992). Through AR functionalities in the CO3 mobile application, participants to an ACA can discover 3D augmented reality objects that are linked to social networking, deliberation and tokenisation activities in the other platform components (FirstLife, LiquidFeedback, the wallet). AR markers (see Figure 6.3) anchor the AR 3D objects to physical environments. AR markers are created specially for each ACA and are positioned in its physical spaces (e.g. hung on a public space or on a table). When people in the ACA scan the AR marker with their CO3 AR app, the CO3 mobile app reveals the AR objects present in the environment, allowing the user to interact through them and participate in related activities. Note that we used AR markers and not smartphone GPS to geolocate the user. This is because GPS can be highly inaccurate for urban spaces such as an association's courtyard. Furthermore, the project must also support the presence of social activities in indoor locations, where GPS performance degrades significantly.

Since AR technology is not available on all smartphones, CO3 offers alternatives to 3D view for finding digital objects related to the ACA. Other proximity-related data visualisations are: a 2D view on the FirstLife geolocated map, and a 1D view consisting of a list of entities in the FirstLife wall sorted by distance or importance. This approach makes it possible to expand the audience of users able to use the CO3 mobile application. In accordance with the concept of ACA, AR contents can be created, discovered and interacted with when the user is physically present (and can be geolocated) in the proximity of the location where the new content is accessible. This way, only people actually living in this

Fig. 6.3 Example of AR Marker. Image created by authors.

urban area are allowed to shape the virtual world connected with it. AR markers are the point of union between the digital and physical reality. Users scan the AR marker through the "augment my city" button in the app, and add onto the AR space 3D objects representing tasks, activities, deliberations, tokens, coupon boxes, crowdsales; or they can tag objects in the physical world with additional digital information. All objects are thus geolocated in the ACA and visualised accordingly on the FirstLife map. Other users will scan the AR markers, discover the objects, and interact with them, for instance by picking up a coupon box and transferring it into their personal wallet, or deciding to join a crowsdale and then sending to it coins held in their wallets. As for the wallet, in-presence interactions are also fostered with QR code mechanisms. Users can send transactions to other users by scanning the QR code associated with their wallet. The original ACA concept, based on in-presence-only interactions was partially relaxed following the constraints imposed by the COVID-19 pandemic and other pilot-specific requirements in order to make it possible, under specific conditions, to add content remotely, provided that the users are subscribed members of the ACA.

3. Economic Urban Commoning

The concept of augmented commoning has been contextualised in local service scenarios covering different areas of applications: social care and solidarity, participatory urban planning, and economic commoning. Blockchain-enabled transactions are most relevant to economic commoning as it is the core of augmented socio-economic networking processes. In this section, we present different economic commoning scenarios as implemented in local pilots in the cities of Turin, Italy (augmented neighbourhood houses) and Athens, Greece (Groceries on-Hold). The former is presented in detail, and focuses on socio-cultural and commercial activities at the neighbourhood level. The latter provides further examples focused on solidarity and donation practices.

3.1 Methodology

The CO3 pilots were implemented based on the Urban Living Labs approach. Living Labs are experimental ecosystems for creating and testing new services and technologies based on active user involvement, co-creation approach, and multi-stakeholder participation (Ballon & Schuurman, 2015; Schuurman & Tõnurist, 2017; Hossain et al., 2019). Urban Living Labs pays specific attention to geographical situatedness in local socio-political contexts, intentional ongoing learning, partnership between local authorities and research institutions, and the active role of citizens (Aernouts et al., 2023; Bulkeley et al., 2019).

The definition of high-level requirements for both the collaborative services at stake and their enabling technologies was based on co-design methodologies (Pautasso et al., 2021). The methodologies used were based on service design (Polaine et al., 2013), user centred design (IDEO, 2015), participatory approaches (Sanders et al., 2010; Stepanek, 2015; Teli et al., 2020), and related methods and techniques. The co-design process first focused on possible use cases, which together constitute an overarching scenario defined as economic urban commoning. Then, more specific requirements for the app functionalities have been identified. Based on this, the development of the core technical features of the CO3 mobile app (see Section 2) was complemented by pilot-specific and scenario-specific customizations. Participants to the pilots took part in iterative testing sessions of the app prototype and of its improved versions. Through focus group techniques, semi-structured interviews and questionnaires, the research team gathered the participants' feedback on the following topics: overall relevance of the digitally-enabled service for daily practices; usability of the integrated app and of specific functionalities; suggestions for UX and UI improvements; sustainability of the service in the local contexts.

3.2 Augmented Neighbourhood Houses in Turin

In the Turin pilot, researchers and local actors co-designed a single broad service scenario exploring different aspects of economic commoning in augmented neighbourhood houses. Neighbourhood houses are social and cultural laboratories that facilitate participation, involvement and self-organisation. Nine such houses have been created thanks to innovative urban regeneration policies developed by the City of Turin municipality since the late 1990s. The pilot scheme focused on three such houses. The three neighbourhood houses involved in the project tested tokenised economies for incentivising and enhancing the exchange of economic and social values within the house economies. Tokens were used to represent both monetary values (e.g. in the form of local coins or discount coupons) and non-monetary values (e.g. volunteering time, which is made more liquid with tokenisation). Coins, coupons, other kinds of interoperable tokens, coupon boxes, and crowdsales allowed the implementation of core mechanisms at the basis of the economic commoning as follows: First, the associations that run the neighbourhood houses created their own virtual currency in the form of local coins to incentivise exchanges within the community. Citizens purchased the coins as prepaid cards, and used them to purchase goods or services. Other social and economic actors within the house's ecosystem created coupons representing different kinds of loyalty tools which are interoperable with the local coin. Second, the CO3 wallet was also used as an accounting system to record users' contribution to life in the neighbourhood houses. Time spent performing voluntary work pertaining to common spaces in the neighbourhood houses were accounted for, tokenised and converted into local credit through coins and coupons.

Thanks to *augmented reality* functionalities integrated with the wallet in the CO3 app, citizens who entered the local ACA (the physical house and its surroundings) were made aware of the opportunity for economic commoning. For instance, by scanning the AR markers that were present on the houses' physical walls and other structures, they could visualise and interact with a variety of contents related to ongoing and upcoming activities taking place in the neighbourhood house, including coupon boxes offering gift coupons and notifications about associations and shops providing discount coupons, open crowdsales, volunteering tasks, or cultural and social events (see Figures 6.4 and 6.5). The same information could also be found on the FirstLife map of the ACA.

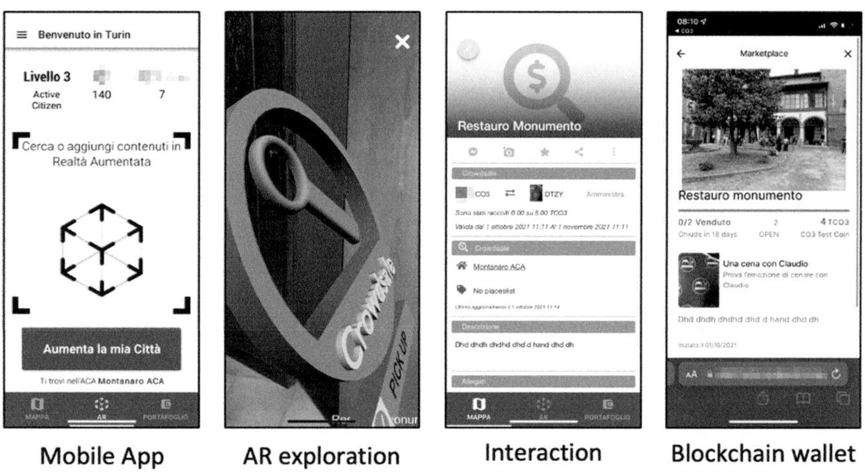

Fig. 6.4 CO3 mobile app: AR, interaction and blockchain wallet functionalities. Image created by authors.

Fig. 6.5 CO3 mobile app: Blockchain wallet functionalities: wallet, marketplace, crowdsale. Image created by authors.

The *gamification* module developed for the Turin pilot scheme provided a further set of incentives for active participation. It included individual badges (medals that every user could receive according to the activities they performed), periodic badges (medals that acknowledged different goals if reached in a given time slot) and collaborative badges. Collaborative badges were designed to enhance the collective dimension of the commoning and resulted from the activities of the whole community. They were assigned to every user as a member of the commons, regardless of their skill or contribution. Taken together, different degrees of participation were recognised from the Rookie level of common citizens, to Fellow, Co-creator and Commoner (see Figure 6.6).

The use cases below showcase how the ACA works in practice.

Use Case 1: *Engagement in the ACA with blockchain and gamification.* The neighbourhood house creates its own coin with the aim of enabling and facilitating the house's visitors to use its services and participate in the commons. A theatre company prepares a coupon to invite people to subscribe to its acting course. Monica, a visitor to the ACA, discovers the coupon for the course thanks to AR. Having read the details, she decides to subscribe. Through the wallet, Monica books the course using the house coins. After payment, she receives a reward (a token, point or badge) for using the app. Motivated by this gamification strategy, Monica decides to buy and drop a "service on-hold" (a coffee, a meal, a music lesson…) to be offered to someone in need.

Use Case 2: *Volunteer management with blockchain and gamification.* Volunteers are invited to download the CO3 app. Management staff prepare a list of activities useful/needed for the neighbourhood house. Each activity is assigned a certain

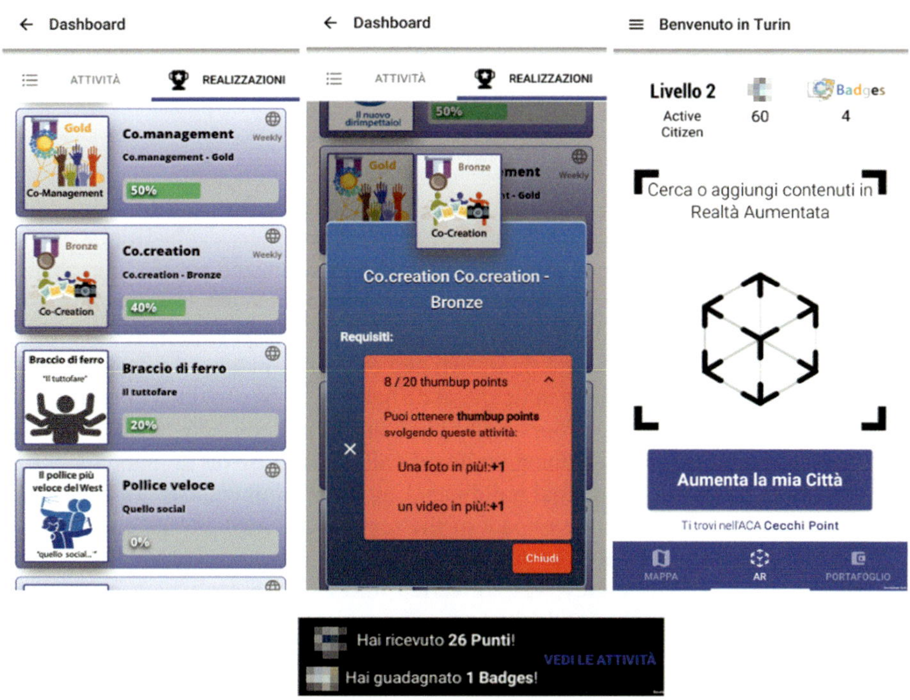

Fig. 6.6 CO3 app: gamification badges and overview of the user's achievements. Image created by authors.

time duration, a certain amount of fungible tokens as a reward, and/or a unique token that certifies who completed the task. The fungible tokens can be used to get discounts for services offered by the neighbourhood house. Louis, the manager, then logs in to the app and augmented area and drops objects representing the tasks. At a later point, he checks which tasks have been collected in the AR area and verifies that they have been completed in the real world before then assigning the rewards. The time spent in volunteering is converted into different kinds of tokens (tokens provided by a specific association, or by the neighbourhood house's managing body). Sara, a volunteer, is able to use the tokens gained through her activity via dedicated smart contracts.

Use Case 3: *Volunteer management with AR.* Andrei is a gardener in a public park in Turin. He checks a new plantation of flowers that a local volunteering group of students was supposed to do. He accesses the ACA through his phone and scans the area of the park in his sight. There, he can see highlighted the area where new flowers have been planted. There is also a 3D object symbolising the relevant action, placed there by the local volunteering group. By clicking on the object, he can see the decision-making process in LiquidFeedback. Students and park users have discussed and decided which type of flowers to plant and how to finance the project. The project had been financed by a token the student group had initiated through the blockchain application. This token was bought by other users, who in return got tokens for discounted meals from local restaurants.

3.3 Groceries On-Hold in Athens

The pilot implemented in Athens provides an example of economic commoning more focused on donation and solidarity practices. The Groceries On-Hold scenario, tested in two urban farmers' markets, allowed disadvantaged people to exchange donated digital tokens for fresh, quality food. The pilot scheme was inspired by the popular Coffee On-Hold initiative (also known as 'suspended coffees'[5]) where customers could, when buying their own cup of coffee, make a donation towards the coffee of other customers in financial difficulties to be provided upon request. With Groceries On-Hold, donors could put coins in various designated electronic internet-connected wooden donation boxes called Coop Boxes at market stalls in farmers' markets and municipal buildings. Using AR Markers, donors could later obtain information about the impact of the service and gain gamification points. The Coop Boxes were able to automatically count the coins inserted through a micro-computer. The money was converted into CO3 tokens which were given to beneficiaries – people who had a home but struggled to feed their families – so that they could obtain free vegetables. A QR code was placed on the stall of every participating vendor which operated as a public address of their digital wallet. Beneficiaries scanned this QR code with their CO3 digital wallet to consume their tokens. Vendors were compensated with cash collected in Coop Boxes in accordance with the tokens consumed at their stalls.

4. Potentials of ACAs and Tokenised Commoning

From a technical point of view, the CO3 app successfully integrated several technologies to allow proximity-based blockchain transactions, content creation, democratic feedback, and gamification-based rewards. The co-designed service scenarios showcased the potential of augmented commoning areas to revolutionise interactions between citizens and public administrations. The blockchain-based tokenised economies were strongly related to specific urban spaces through geolocation functionalities and AR-enabled discovery. Digitally-enhanced socio-economic interactions was made accessible through three different entry points: AR, a geolocated map and proximity-based listing. This broadened our conception of augmented commoning beyond augmented reality, with 2D (map-based) and 1D (listings- based) views. All this brings participants to the augmented commoning areas, to be physically present in these public urban spaces, and to meet each other in person. This helps to foster attendance in places that are socially, culturally and economically significant in the neighbourhood or city, and to encourage personal encounters.

---

[5] See for instance https://suspendedcoffees.com

The pilots allowed us to explore the feasibility of ACAs and to identify their strengths and weaknesses. On the one hand, it was clear that urban commoners, local public officers and social workers are interested in using innovative techniques for fostering co-production processes, and we verified that the integration of complementary disruptive technologies within an app for urban commoning is technically feasible. On the other hand, the actual introduction of these technologies in real-life contexts proved to be quite challenging. Some criticalities related to lack of familiarity with AR and phygital interactions, or lack of devices with the required AR functionalities, resulting in limited entry points for some users. During the design phase, this aspect was taken into account; in fact, 3D objects are used only during the discovery and navigation phase of the environment. Once the artefact is selected, user interaction takes place with a standard interface. Nevertheless, some users experienced a higher degree of effort and less smooth user experience than they expected when using the integrated functionalities. We conclude that the introduction of disruptive technologies in social dynamics and collaborative processes should be an incremental rather than immediate process. It requires financial, legal and cultural preconditions as well as the strong presence of digital facilitators who accompany public officers and citizens in adopting the system. There was also one unexpected challenge that had a significant impact on the rollout of the pilot schemes; the Covid-19 pandemic. Restrictions on gatherings meant that some functionalities intended to be in-presence-only had to be available remotely. The pandemic affected the duration of the pilots as well as the scheduled digital facilitation for participants, similarly to what happened to many research proojects (Foth et al., 2021).

There were also some specific challenges concerning blockchain-based wallets and related tokenised economies. First, co-designing tokenized economies in social, civic and commoning contexts requires careful understanding of which interactions can be tokenized with positive effects on the desired community goals, and which ones would bring the risk of an excess of quantification and commodification. Attention must be paid to ensuring that tokenized incentives do not simplify social relationships or crowd out intrinsic or altruistic motivations, especially when social values, and not only material and monetary values, are tokenized. Similarly, it is necessary to distinguish whether some processes can be made more efficient and effective through smart contracts, or there is a risk of an excess of automation of mutable social dynamics. Secondly, there is currently a lack of legal and administrative frameworks regulating blockchain token economies (e.g. the sale of tokens as local coins). Therefore, compromises were needed for administering transactions of tokens by adapting an ordinary accounting system. Third, end users reported some distrust of donation systems (Athens) or prepaid services (Turin) that required putting money into an experimental system. Fourth, tokens and tokenised economies are not well-known concepts and mechanisms yet. Moreover, the user experience related to some aspects of the blockchain wallet was not fully intuitive. The interaction models co-designed with some of the local actors involved would have needed further adaptations in order to, for instance, comply with the different needs of particular neighbourhood communities in Turin. However, we encountered interest on the part of active citizens, association managers, commoners and public officers in how the blockchain can make commoning and community economies more vital and engaged.

These lessons have prompted us to continue experimenting with blockchain-based solutions for local communities, building on the experience of the CO3 pilots. In particular, the following criteria guide our new experimentations: First, blockchain-based wallets are designed with an interface that is as intuitive as possible for end users. Secondly, modular and flexible functionalities are designed within the wallet so that they are easily adaptable to different tokenised economic models in different local communities. Third, we continue the CO3 work of integrating functionalities for tokenisation with geolocation and proximity interactions. All these considerations have led us to work on the wallet app CommonsHood (Balbo et al., 2020): a new decentralised application specifically oriented towards enabling social collaborative economies and financial inclusion processes in local communities. CommonsHood enables any user (individual or organisation) in a local community to create cryptographic tokens for their own purposes from a set of predefined smart contract templates, using a simple web app interface developed according to a no-code approach. Tokens can represent assets of value such as customer loyalty instruments, prepaid cards, complementary currencies, digital collectibles, sharing rights (e.g. to a tool or infrastructure), digitalised representations of material entities (e.g. a ticket to an event). Or, they can be purpose-driven tokens to incentivise individual behaviours towards common goals. Starting with the definition of the concept of Web2.0, we define this approach as civic blockchain (Viano et al., 2023). Like the CO3 wallet, CommonsHood also allows users to create crowdsales. Furthermore, new functionalities have been added: the creation and management of NFTs through a marketplace, exchange of tokens, and more complex functionalities such as a "library of things" that allows the decentralised management of sharing economies in local communities (e.g. loaning objects between neighbours). In continuity with the ACA concept, geolocation and proximity interactions are enabled via integration with the FirstLife map, and by allowing the transfer of tokens between wallets via QR code. The definition of the CommonsHood concept and functionalities began with co-design activities implemented in the framework of former European projects [examples anonymised] and the output of CO3. Co-design methodologies were used to define high-level requirements for both collaborative services and their enabling technologies (see Section 3.1) Each experiment always began with a preliminary co-design phase that allowed the platform to be enriched over time, making it a valuable tool capable of being used in many real-world scenarios.

Conclusions and Future Work

This chapter described the development of augmented commoning in the context of CO3, a project that addressed innovation of public services co-production with blockchain and other disruptive technologies. The project pertains to the domain of citizens' participation in the civic, democratic and economic life of their local community. In this domain, blockchains have not yet been significantly explored. The CO3 project and this chapter aimed at contributing to explore the socio-cultural, political and technical implications of adopting blockchains at the local community level. CO3 showed that tokenisation, smart contracts and tokenised economies can provide common and interoperable tools as a useful layer to represent community rules and create new intercommunity services. The CO3 app integrated disruptive technologies to augment connections between people in the same place and allow proximity-based blockchain transactions. Indeed, the CO3 ACA model can be regarded as a first step towards effective methodologies for co-designing and implementing blockchain-based tools that are ethical and desirable for enabling the co-production of public services in local communities.

CO3 showed the possibilities of augmented urban commoning areas, where new technologies go beyond mediation with interactive technologies (e.g. social media, online collaboration platforms) and enables us to re-evaluate spatially-intermediate interactions. It also showed the possibilities of exploiting disruptive technologies for collaborative purposes in the public and civic realm. CO3 introduced ambitious, innovative, community-based pilot schemes which allowed us to gain understanding of the kind of technical, administrative and social challenges that can arise when implementing ACAs. It also highlighted that technologies alone do not guarantee that transformative changes happen if they are not conceived of as part of a broader process of innovation that is not only technical but also creates social, cultural, economic, and legal framework preconditions. Moreover, the presence of digital facilitators to accompany these processes is crucial. Some preconditions are not yet widespread, for example knowledge of interactive democracy, availability of mobile devices with advanced AR features, and legal frameworks regulating blockchain token economies. These considerations have oriented our subsequent research on blockchain for civic purposes. At present, the concept and functionalities of the CommonsHood wallet app build on the lessons learned in CO3 to further improve the blockchain wallet's usability and its adaptability to different local contexts. Future work will include: a study of the legal and administrative framework that can frame token economies, the development of new functionalities, and further co-design of new service scenarios that can be enabled through the app.

In conclusion, we believe that augmented commoning has the potential to revolutionise interactions between citizens and public administrations. The CO3 project has demonstrated that a range of complementary, advanced technologies can be integrated in a community-based blockchain app for this purpose. Collaborative projects that focus on the societal as well as the technical hold promise for a bright future of immersive, engaging and fulfilling co-creation, co-production and co-management of public spaces.

Acknowledgements

This work has received financial support from the European Union's Horizon 2020 research and innovation programme 2019-2021 under grant agreement No 822615 (project "CO3: Digital Disruptive Technologies to Co-create, Co-produce and Co-manage Open Public Services along with Citizens"), and from ICSC – Centro Nazionale di Ricerca in High Performance Computing, Big Data and Quantum Computing, funded by European Union – NextGenerationEU.

References

Aernouts, N, Cognetti, F, and Maranghi, E (2023). *Urban living lab for local regeneration* In: Beyond Participation in Large-scale Social Housing Estates. Cham: Springer.

Bagloee, S.A., Heshmati, M., Dia, H., Ghaderi, H., Pettit, C. and Asadi, M. (2021). Blockchain: The operating system of smart cities. *Cities*, *112*,103104.

Balbo, S., Boella, G., Busacchi, P., Cordero, A., De Carne, L., Di Caro, D., Guffanti, A., Mioli, M., Sanino, A. and Schifanella, C., (2020). CommonsHood: A blockchain-Based Wallet App for Local Communities. Presented at the Proceedings - 2020 IEEE International Conference on Decentralized Applications and Infrastructures, DAPPS 2020, pp. 139–44.

Ballon, P. and Schuurman, D. (2015). *Living Labs: Concepts, Tools and Cases*, vol. 17. London: Emerald Group Publishing Limited.

Behrens, J., Kistner, A., Nitsche, A. and Swierczek, B. (2014). *The Principles of LiquidFeedback*. Berlin, Germany: Interaktive Demokratie.

Bollier, D., Helfrich, S. (2019). *Free, fair, and alive: The insurgent power of the commons*. New Society Publishers.

Boyte, H.C. (2004). *Everyday Politics: Reconnecting Citizens and Public Life*. University of Pennsylvania Press. http://www.jstor.org/stable/j.ctt3fj16v.

Brandsen, T. and Honingh, M. (2018). Definitions of co-production and co-creation. In Brandsen, T., Verschuere, B., Steen, T. (Eds.). *Co-Production and Co-Creation. Engaging Citizens in Public Services, New York*, (pp. 9–17). Routledge. https://doi.org/10.4324/9781315204956.

Bulkeley, H., Marvin, S., Palgan, Y.V., McCormick, K., Breitfuss-Loidl, M., Mai, L., et al. (2019). Urban living laboratories: conducting the experimental city? *Eur Urban Reg Stud, 26*, 317–35. 10.1177/0969776418787222.

Certomà, C. (2021). *Digital Social Innovation: Spatial Imaginaries and Technological Resistances in Urban Governance*. Springer International Publishing.

Cila, N., Ferri, G., de Waal, M., Gloerich, I. and Karpinski, T. (2020). The blockchain and the commons: Dilemmas in the design of local platforms. *Proceedings of the 2020 CHI Conference on Human Factors in Computing Systems* (pp. 1–14) [3376660]. ACM-SIGCHI.

Domenicale, I., Viano, C. and Schifanella, C. (2024), Blockchain for local communities: an exploratory review of token economy aspects. *Front. Blockchain, 7*, 1426802.

Estrada-Grajales, C., Foth, M. and Mitchell, P. (2018). Urban imaginaries of co-creating the city: Local activism meets citizen peer-production. *Journal of Peer Production*, 11.

Fines Schlumberger, J-A. (2022). Blockchains & Sustainable Development Report, Association de loi 1901 Blockchain for Good - France, blockchainforgood.fr September 2022.

Foster, S. and Iaione, C. (2016). The city as a commons. 34 *Yale L. & Pol'y Rev*. 281 http://dx.doi.org/10.2139/ssrn.2653084.

Galen, D., Brand, N., Boucherle, L., Davis, R., Do, N., El-Baz, D., Kimura, I., Wharton, K. and Lee, J. (2018). *Blockchain for Social Impact: Moving Beyond the Hype*. Stanford, Center for Social Innovation, RippleWorks.

Hossain, M., Leminen, S. and Westerlund, M. (2019). *A systematic review of living lab literature*. J Clean Prod.

IDEO, (2015). *The Field Guide to Human-centered Design: Design Kit*. IDEO.org. https://www.ideo.com/journal/design-kit-the-human-centered-design-toolkit.

Lember, V. (2018). The increasing role of digital technologies in co-production and co-creation. In Brandsen, T., Verschuere, B., Steen, T. (Eds.), *Co-Production and Co-Creation. Engaging Citizens in Public Services. New York*, (pp. 115–27). Routledge.

Lewis, J.M., McGann, M. and Blomkamp, E. (2020). When design meets power: design thinking, public sector innovation and the politics of policymaking. *Policy & Politics*, *48*(1), 111–130.

Mattsson, C.E.S., Criscione, T. and Takes, F.W. Circulation of a digital community currency. *Sci Rep* 13, 5864 (2023).

Miller, E. (2010). Solidarity Economy: Key Concepts and Issues. In Kawano, E., Masterson, T., Teller-Ellsberg, J. (Eds.), *Solidarity Economy I: Building Alternatives for People and Planet*. Amherst, MA: Center for Popular Economics.

Monno, V. and Khakee, A. (2012). Tokenism or Political Activism? Some Reflections on Participatory Planning. *International Planning Studies*, *17*(1), 85–101.

Murray-Rust, D., Elsden, C., Nissen, B., Tallyn, E., Pschetz, L. and Speed, C. (2022). Blockchain and Beyond: Understanding Blockchains through Prototypes and Public Engagement. *ACM Trans. Comput.-Hum. Interact*. 29, 5, Article 41 (October 2022), 73.

Nissen, B., Pschetz, L., Murray-Rust, D., Mehrpouya, H., Oosthuizen, S. and Speed, C. (2018). GeoCoin: Supporting Ideation and Collaborative Design with Smart Contracts. *Proceedings of the 2018 CHI Conference on Human Factors in Computing Systems*, 1–10.

Olivier, P. and Wright, P. (2015). Digital civics. *Interactions*, *22*, 61–63.

Ostrom, E. (1990). *Governing the commons: the evolution of institutions for collective action*. United Kingdom Cambridge University Press.

Polaine, A., Løvlie, L. and Reason, B. (2013). *Service design: From insight to implementation*. Rosenfeld media.

Pólvora, A., Hakami, A., Bol, E. (Eds). (2020). *Scanning the European Ecosystem of Distributed Ledger Technologies for Social and Public Good: What, Why, Where, How, and Ways to Move Forward*. EUR 30364 EN, Publications Office of the European Union, Luxembourg, ISBN 978-92-76-21578-3.

Pautasso, E., Frisiello, A., Chiesa, M., Ferro, E., Dominici, F., Tsardanidis, G., Efthymiou, I., Zgeras, G. and Vlachokyriakos, V. (2021). The Outreach of Participatory Methods in Smart Cities, From the Co-Design of Public Services to the Evaluation. *International Journal of Urban Planning and Smart Cities (IJUPSC) 2*(1).

Rehm, S., Foth, M. and Mitchell, P. (2018). DoGood: examining gamification, civic engagement, and collective intelligence. *AI and Society*, 33(1), 27–37.

Robles, A.G., Hirvikoski, T., Schuurman, D. and Stokes, L. (2016). *Introducing ENoLL and its living lab community*. ©ENoLL Available at: https://issuu.com/enoll/docs/enoll-print.

Rodrigo, L., Palacios, M. and Ortiz-Marcos, I. (2019). Digital Social Innovation: analysis of the conceptualization process and definition proposal. *Dirección y Organización*, 59-66. 10.37610/dyo.v0i67.545.

Sanders, E.B.N., Brandt, E. and Binder, T. (2010). A framework for organizing the tools and techniques of participatory design. In *Proceedings of the 11th biennial participatory design conference*, pages 195–98.

Schuurman, D. and Tõnurist, P. (2017). Innovation in the Public Sector: Exploring the Characteristics and Potential of Living Labs and Innovation Labs. *Technol Innov Manag Rev* 7, 7–14. 10.22215/timreview/1045.

Shen, C. and Pena-Mora, F. (2018). Blockchain for Cities—A Systematic Literature Review. *IEEE Access*, *6*, 76787–76819.

Stepanek, M. (2015). Building with, not for. *Stanford Social Innovation Review*: June, 9:2015.

Tallyn, E., Revans, J., Morgan, E. and Murray-Rust, D. (2020, July). GeoPact: Engaging publics in location-aware smart contracts through technological assemblies. *Proceedings of the 2020 ACM Designing Interactive Systems Conference* (pp. 799-811).

Teli, M., Foth, M., Sciannamblo, M., Anastasiu, I. and Lyle, P. (2020). Tales of Institutioning and Commoning: Participatory Design Processes with a Strategic and Tactical Perspective. *Proceedings of the 16th Participatory Design Conference* (PDC 2020), 159–71.

Viano C., Avanzo S. and Boella G. Schifanella C. (2023). Civic Blockchain: Making blockchains accessible for social collaborative economies. *Journal of Responsible Technology, 15*, 2023.100066.

Vlachokyriakos, V., Crivellaro, C., Wright, P., Karamagioli, E., Staiou, E.-R., Gouscos, D., Thorpe, R., Krüger, A., Schöning, J., Jones, M., Lawson, S. and Olivier, P. (2017). HCI, Solidarity Movements and the Solidarity Economy. *Proceedings of the 2017 CHI Conference on Human Factors in Computing Systems*, 3126–37.

Vlachokyriakos, V., Crivellaro, C., Le Dantec, C. A., Gordon, E., Wright, P. and Olivier, P. (2016). Digital Civics: Citizen Empowerment With and Through Technology. *Proceedings of the 2016 CHI Conference Extended Abstracts on Human Factors in Computing Systems*, 1096–99. ACM.

Voshmgir, S., Wildenberg, M., Rammel, C. and Novakovic, T. (2019). *Sustainable Development Report: Blockchain, the Web3 & the SDGs*. Research Institute for Cryptoeconomics.

Xie, J., Tang, H., Huang, T., Yu, F. R., Xie, R., Liu, J. and Liu, Y. (2019). A Survey of Blockchain Technology Applied to Smart Cities: Research Issues and Challenges. *IEEE Communications Surveys and Tutorials, 21*(3), 2794–2830.

Yuan, Q. (2019). Co-production of Public Service and Information Technology: A Literature Review. *Proceedings of the 20th Annual International Conference on Digital Government Research*, 123–132.

7 A Reference Architecture for Blockchain-based Social Good Applications

Gowri Sankar Ramachandran[1*] and Raja Jurdak[2]

1. Introduction

Blockchain technology has revolutionised the financial industry, with many governments considering blockchain-based digital currencies (Alam et al., 2021) or Central Bank Digital Currency (Auer et al., 2022). Additionally, several non-cryptocurrency applications have been explored in the supply chain (Ramachandran et al., 2022), real estate (Garcia-Teruel, 2020), healthcare (Garcia et al., 2022), finance (Natanelov et al., 2022), energy (Jurdak et al., 2021), and data marketplaces (Ramachandran et al., 2018). These applications highlight the practical significance of blockchain technology in several application sectors. The decentralisation of application processes is one of the key drivers for such widespread adoption.

Before the arrival of blockchain technology, applications were developed with distributed architectures, wherein the technical functionalities were handled by computation devices belonging to a single organisation. As the application user, the distributed architectures may appear decentralised. However, the infrastructure is under the control of a single entity, introducing a single point of failure. One may argue that the practical byzantine fault-tolerant algorithm offered a mechanism for running distributed applications involving multiple organisations (Castro et al., 2002), but it was not popular and widely adopted prior to the emergence of blockchain technology.

The shift to *decentralised architectures* from *distributed architectures* has become a common theme since the introduction of BitCoin in 2008 (Nakamoto, 2008). Since then, many researchers and practitioners applied blockchain technology to create decentralised applications with transparency and auditability. It is important to underscore that the information stored on the blockchain is visible to anyone. This means anyone can view the BitCoin blockchain ledger to understand the currency transactions between accounts. However, the anonymity of BitCoin protects the privacy of BitCoin owners. Despite the privacy restrictions, blockchain technology enables application developers to maximise accountability in multi-stakeholder applications such as supply chains.

[1,2] Queensland University of Technology, Brisbane, Australia
Email: r.jurdak@qut.edu.au[2]
* Corresponding author: g.ramachandran@qut.edu.au

Properties such as decentralisation, transparency, and auditability make the blockchain a suitable technology for applications with social benefits. Existing literature presents the application of blockchain to diamond tracing (Cartier et al., 2018), farm-to-plate tracking (Cao et al., 2021; Lan et al., 2017), and provenance for coffee beans (Tharatipyakul et al., 2022). These applications aimed to disintermediate the supply chain and incentivise the farmers by removing the middlemen, maximising the profit for the producers rather than distributors. Thus, blockchain technology has the potential to benefit farmers and other less visible entities. However, building useful blockchain applications is challenging due to a diversity in platforms and functionalities, such as the need to support legacy applications and AI for predictions.

This chapter proposes the design and development of a reference framework for blockchain applications, focusing on social good applications and presents an early blueprint of this framework. It starts with the requirements of blockchain applications targeting social good. Subsequently, it introduces a set of technical frameworks and a reference architecture to help practitioners and application developers build impactful blockchain-based applications for a fairer and just society.

2. Blockchain's Key Properties for Social Good Applications

Blockchain-based decentralised applications are also called Web3 because they promise to deliver a new version of the Internet powered by blockchain technology (Stackpole, 2022). Web2, on the other hand, allows people to build applications following a client-server technology, wherein the users act as clients and interact with a remote server under the control of a third party. While Web2 revolutionised society in many ways, it also had its fair share of controversies.

The Facebook-Cambridge Analytica scandal in 2020 revealed that users' sensitive data may be harvested and utilised for election campaigns through personalised advertisements (Isaak et al., 2018). One may argue that such a scandal happened because of people's lack of awareness of data privacy and good data practices, but it also underscores the limitations of centralised architectures that are at the forefront of Web2. Each user maintains an account on a platform that is under the control of an organisation such as Facebook. In this setting, the organisation that manages the infrastructure has complete control over the user's data unless the user makes an effort to protect sensitive data. Therefore, the centralised architectures found in Web2 are inadequate for transparency and accountability. Web3 promises to provide more transparency to the users by leveraging the blockchain technology at its core.

3. Building Blocks of Blockchain

The blockchain technology at its core includes cryptographic algorithms, consensus protocols, peer-to-peer networking, immutable storage, and smart contracts. This section provides an overview of such building blocks and explains how they make blockchain a perfect platform for social good applications.

3.1 Public-key Cryptography

The public key cryptography uses a key pair: a public key and a private key. It is useful for digitally signing transactions in an untrusted environment. In blockchain-based applications, a user can prove the authenticity of a transaction by signing the transaction using the private key. When a recipient of the transaction wants to verify its authenticity, they can use the public key to verify the ownership, thus ensuring authenticity. Additionally, blockchain-based applications don't use dedicated identity management solutions to track users within the application network. Instead, the user's public key is one of the ways in which the user can be identified. However, it does not reveal any sensitive information, such as the name, date of birth, or location. Therefore, users carry out transactions anonymously in blockchain-based applications. Some literature argues that the use of public keys to identify users on the blockchain violates the anonymity property. Hence, it provides pseudonymity, meaning the user can be identified with their public key but not with their real name or other personally identifiable information (Andola et al., 2021). In summary, public key cryptography allows users to carry out transactions without disclosing their true identity, while still having mechanisms to prove transactions' authenticity.

Application developers and users benefit from public key cryptography, which provides privacy by hiding the real identities of the stakeholders while still offering mechanisms to authenticate and verify the transactions. Social-good applications can use this property to verify and question the credibility of transactions when there is a discrepancy.

3.2 Consensus Algorithm

Unlike a centralised system, the blockchain operates in a distributed setting, wherein the transactions are processed by a set of globally distributed nodes. The consensus algorithm, which runs on each blockchain node, collectively approves and finalises the transactions as part of the block creation process. Some of the popular consensus algorithms include the Proof-of-Work with the longest chain rule (Nakamoto, 2008), Proof-of-Stake (Saleh, 2021) and the practical Byzantine Fault Tolerant (PBFT) (Castro et al., 2002).

The BitCoin blockchain, for example, uses the Proof-of-Work algorithm to select a special leader node every 10 minutes through a rigorous mining process involving a cryptographic puzzle. The winner of the puzzle challenge in each round gets to propose the next block to be added to the blockchain. All the nodes in the BitCoin network receive the proposed block from the successful miner and then add them to the blockchain ledger after verifying the block's integrity. Since it is a distributed protocol, more than one node may have solved the puzzle at the same time to become a leader, which would lead to multiple blocks being proposed at the same time. In such cases, the longest chain rule is applied by the BitCoin nodes, wherein the longest blockchain in the network is collectively accepted as the credible ledger following a distributed consensus process. In summary, the BitCoin ledger follows a distributed and decentralised consensus process to handle network transactions without relying on a single node or authority, unlike a centralised system.

The PBFT is another consensus protocol that was proposed in the late 90s. It also operates in a distributed network but expects the majority of the nodes in the network to approve a given transaction for it to be considered valid. More specifically, each transaction or block must be approved by more than two-thirds of the nodes in the network following this consensus protocol. For instance, in a blockchain application with ten nodes, each transaction must be approved by at least seven or more nodes. The protocol operates by exchanging transactions and their approvals among each other before counting the tally of successful approvals. Upon successful consensus, the blockchain is updated with the new block. One of the key differences between PBFT and Proof-of-Work, which has the longest chain, is that PBFT does not include cryptographic puzzle challenges, which makes it efficient for decentralised applications. Tendermint (Buchman, 2016) and Hyperledger Besu (Fan et al., 2022) are examples of PBFT-based blockchains.

This feature is fundamental for social-good applications as it ensures that the key application processes are not controlled by a single entity. By having consensus nodes from each stakeholder, applications can operate in a democratic environment. Different variants of consensus protocols can be employed, such as the social consensus for the supply chain (Powell et al., 2021), to strengthen the credibility and integrity of social-good applications.

3.3 Immutable Ledger

Centralised applications typically store application transactions in a database. Since the application infrastructure is under the control of a single entity, the entries in the database could be manipulated or modified without any transparency or accountability. The blockchain technology overcomes this problem by storing transactions in a distributed and immutable ledger. Simply put, the blockchain transactions are collectively stored in all the nodes in the network after a successful consensus process. When a block is added to the ledger after successful consensus, it cannot be modified. When the data gets stored in a centralised database, it may be easy to modify one or more entries without offering any transparency. However, the blockchain ledger is stored and managed by tens to hundreds of nodes, wherein changing an entry in the block is almost impossible due to the computational complexity and the inherent security features of blockchain protocols. Because of this reason, the immutable ledger is often termed a *write-once* ledger, as applications can only write data once in the ledger, which can never be modified or rewritten, offering transparency and auditability.

This feature is great for social-good applications as the information cannot be modified by any single entity once it is entered into the ledger, preventing any form of data tampering. However, the application developers and users must ensure that the right information is entered into the ledger, as blockchain cannot distinguish between good data and bad data, which is termed the Garbage-in-Garbage-out problem (Powell et al., 2022).

3.4 Smart Contracts

Web2 applications rely on a remote centralised server to perform computations using inputs provided by clients. In this model, there is no credible mechanism to validate the correctness of the computations. Smart contracts overcome this issue in blockchain by executing computations in a decentralised network of nodes as part of the consensus process. Ethereum (Antonopoulos and Wood, 2018) is a blockchain platform with built-in support for smart contracts. Application developers deploy the smart contracts on the underlying blockchain by defining functions that are triggered through user's transactions (Natanelov, Cao et al., 2022). Unlike centralised computation infrastructure, blockchain platforms execute the functions in smart contracts in an open and transparent manner, enabling transparency and auditability. Users can verify and validate the history of computations by checking the blockchain ledger. Several social good applications (Trotter et al., 2020; Tang et al., 2022; Kahya et al., 2021) leverage smart contracts to automate business processes, reward incentives, and manage supply chains. It is one of the powerful features of the blockchain platform.

4. Key Properties of Blockchain for Social Good

Real-world applications that aim to benefit society require systems that offer transparency and accountability. The following properties of blockchain make it a great platform for such applications:

Decentralisation: Web2 systems operate following a client-server architecture, offering full administrative power to the entity managing the server, thus preventing any accountability. Web3, built on top of blockchain technology, relies on a distributed infrastructure with a decentralised administration model, enabling multiple entities to oversee the operations and the decisions. Social good applications would immensely benefit from such an architecture as they require transparency and accountability.

Tamper-proof Ledger: All information is stored in a database controlled and managed by a collection of nodes in the distributed network in Web3 and blockchain systems. This database only includes entries approved by majority of the nodes in the network, preventing any single party from recording false information. Besides, the database cannot be modified without the approval of a large proportion of nodes, overcoming data tampering. Social good applications require such a database to provide data integrity and trust.

Shared Ownership and Governance: Blockchain infrastructure is not under the ownership of a single organisation. Multiple organisations collectively run the infrastructure, thereby enabling democracy within the application ecosystem. Social good applications require such operational and governance models because of the limited trust and transparency of autocratic systems.

Digital Currency and Tokens: Social good applications requiring payment and incentivisation features can leverage blockchain technology as it provides support for cryptocurrencies and tokens in the form of decentralised applications (DApps). Unlike fiat currency, cryptocurrencies and tokens operate on top of a decentralised infrastructure with a transparent ledger, enabling users and organisations to track the flow of payments. This is crucial for social good applications as no organisation can claim that the money is paid to a particular user without offering any proof.

Decentralised Computation Layer: Blockchain infrastructure provides support for smart contracts, which help application developers automate computation processes in a decentralised system. This increases transparency and accountability for end users. Social good applications can leverage this functionality to run complex business processes within a blockchain infrastructure.

These blockchain properties are well-suited to social good applications because they help the developers build transparent, democratic, and accountable applications while still supporting incentives in the form of tokens or digital currency. In the following sections, we present a few frameworks that leverage these properties.

5. Decentralised Publish-Subscribe Communication and Elimination of Racial and Identity Profiling

Contemporary digital applications deal with a significant amount of data. For example, a supply chain application will include data about the product when it travels between the producers, retailers, sellers, and consumers. By tracking this data, the consumers or other stakeholders can easily understand the status of the product. However, the development of a data-driven application in a multi-stakeholder setting requires a significant amount of trust in the data as well as the infrastructure that handles the data. Here, the movement of data, coupled with the participation of multiple stakeholders, forces the data owners and the application developers to seriously introspect the data flow and the role of each stakeholder in the end-to-end application process. When application data is exchanged or managed by a centralised server, it introduces the possibility of data tampering, as the server administrator has full control over the operational infrastructure. Therefore, having a data communication model with centralised control in a digital application is susceptible to a single point of failure.

Publish-subscribe (pub-sub) and request-reply communication models are prominent in the digital applications in Web2. The majority of the applications are built using one of these communication models to exchange data between entities.

The pub-sub model is adopted when lightweight technology with good operational efficiency is needed. Following the pub-sub model, data providers or owners, known as publishers, connect with data consumers, known as subscribers, in a topic-based interaction framework. In this approach, topics are channels through which data is exchanged, allowing subscribers to receive relevant information by subscribing to specific topics.

To illustrate, consider a scenario where a farmer dispatches his coffee harvest to the manufacturing facility to package it for consumers. Following the topic-based pub-sub communication model, the harvest is uniquely identified with a label like *farmA/localeX/harvestid8Nov2023*. Interested parties or stakeholders, such as other suppliers, distributors, and retailers, can access the data related to this harvest by subscribing to the corresponding topic. By adhering to this design principle, a dedicated topic is created for each shipment, enabling stakeholders to both send (publish) and receive (subscribe to) data seamlessly. This pub-sub model streamlines communication and enhances information dissemination, contributing to more efficient and scalable systems. The simplicity of the model lies in its topic-centric organisation, allowing stakeholders to access data relevant to their interests without unnecessary complexities.

However, contemporary publish-subscribe frameworks, including MQTT (Hunkeler et al., 2008) and Apache Kafka (Garg, 2013), follow a centralised architecture. In this architecture, a single stakeholder runs a centralised data broker and manages the data communication between various parties, introducing a single point of failure. Trinity (Ramachandran et al., 2019) presents a byzantine fault-tolerant publish-subscribe broker to enable multiple stakeholders to share data reliably.

5.1 Trinity: A Decentralised Pub-Sub Broker

The pub-sub broker traditionally includes three key entities: broker, publishers and subscribers. The broker is a central server that coordinates the exchange of messages between the publishers and subscribers. Publishers, as the name suggests, are entities that generate the application data and attach a topic to it before dispatching them to a broker. Upon reception, the broker checks whether there are any subscribers for the topic. If there are active subscribers, the data is relayed to them. Figure 7.1 (left) shows how the centralised pub-sub operates with a broker, publisher, and subscriber. Since this architecture is susceptible to a single failure, a decentralised pub-sub broker is essential.

Trinity (Ramachandran et al., 2019) follows a decentralised architecture, as shown in Figure 7.1 (right). It relies on blockchain technology and operates in a distributed network involving multiple stakeholders. Considering a supply chain application, Trinity (Ramachandran et al., 2019) requires all the stakeholders to run a blockchain node to distribute the administrative power, thereby removing the reliance on a centralised entity. In a nutshell, the messages sent by publishers are not handled by a single broker. Instead, it is collectively processed by a collection of brokers that are connected via the blockchain platform. By connecting the brokers to a blockchain, each message is validated through a consensus process before it gets stored on the blockchain's immutable ledger. Additionally, upon successful consensus and the storage of messages on the immutable blockchain ledger, the messages are sent back to the broker for the consumption of subscribers. It is important to note that the messages that are received by the subscribers include proof that there was a consensus process, and a record of it is available on the blockchain ledger. Thus, publishers and subscribers can audit the blockchain ledger at any time to see the authenticity and integrity of the messages. Conceptually, Trinity guarantees correct message delivery to the subscribers as long as more than two-thirds of blockchain nodes are honest. For brevity, the detailed safety and correctness analysis is not discussed in this chapter, but it can be found in our other publication (Ramachandran et al., 2019).

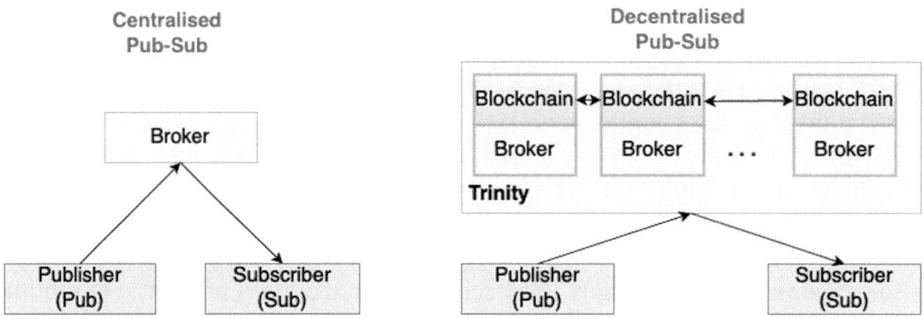

Fig. 7.1 (Left) Centralised Pub-Sub with a single point of failure; (Right) Trinity offers a decentralised Pub-sub with blockchain, overcoming a single point of failure.

5.2 Use Case: Application of Trinity to Prevent Racial Profiling by Law Enforcement

Trinity, by default, requires application stakeholders to jointly operate the application infrastructure. By taking the administrative power from a single entity, Trinity brings in a democratic framework for data communication. With the inclusion of blockchain technology, the data is stored on an immutable ledger, thereby offering reliable and auditable data storage. These features are tailor-made for social good applications, as they require democracy, not autocracy, transparency, not secrecy, and a tamper-proof ledger, not a change-at-will database. As a reason, Trinity was applied as a proof-of-concept to a real-world application.

Trinity was applied for a real-world law enforcement use case. Department of Justice, California introduced a law, the RIPA Act, in 2015 to prevent law enforcement officers from stopping people of ethnicity, colour, and race when they perform traffic stops. In the USA, law enforcement officers perform traffic stops when they suspect people. However, people raised concerns that law enforcement officers are often stopping people of colour, compared to other races. To tackle this issue, a new law called the RIPA Act, was introduced. The goal of the act was to collect traffic stop data from each law enforcement office and law enforcement agency to understand the diversity of people being stopped on the road. This required a data collection framework; however, relying on a centralised infrastructure for this application is susceptible to a single point of failure, as each law enforcement agency can tamper with the data and report only a handful of traffic stop data that demonstrate diversity.

By employing a decentralised data communication and storage protocol such as Trinity, the data collected by law enforcement officers are stored on an immutable ledger after a successful consensus. Note that Trinity is a decentralised and distributed framework requiring the participation of multiple law enforcement agencies, including the Department of Justice. As a result, any entity, including the law enforcement agency or Department of Justice, California, cannot modify the traffic stop data upon successful consensus and storage. Besides, the public can scrutinise traffic stop records on their own to understand the diversity of data. Trinity was employed for this use case with the help of the Los Angeles Sheriff's Department to show its benefit to a social-good use case.

6. Decentralised Request-Reply Framework and its Application to Supply Chain

Trinity overcomes the single-point-of-failure problems in the pub-sub communication protocol. The alternative to pub-sub is the request-reply communication protocol, which dominates the Web2 ecosystem. Unlike pub-sub, the request-reply protocol follows a synchronous messaging model, meaning that the clients who want to access specific data or a resource on a server issue a request and wait for the server's response. Pub-sub, on the other hand, is asynchronous and loosely coupled since the clients simply register their interest in particular data or a resource through the subscription mechanism without waiting for an immediate response. Therefore, the only difference between the pub-sub and the request-reply protocols comes from the way that clients interact with the server or a broker to retrieve the data or a resource. Despite this difference, both protocols follow the client-server architecture, although the server is denoted by the name *broker* in the pub-sub protocol.

As discussed in the previous section on Trinity, a centralised protocol is prone to single-point-of-failure problems. A request-reply protocol includes a centralised server that handles the clients' requests; hence, the server administrator can tamper with the request and manipulate the response without offering any visibility.

Contemporary request-reply frameworks, including web servers, follow the centralised client-server model since Web2 assumes that a single entity is responsible for the administration of the entire operation. However, social good applications demand an application infrastructure with accountability and transparency. Thus, existing centralised frameworks are unsuitable for social good applications. DeWS (Ramachandran et al., 2023) is a decentralised request-reply framework with transparency, auditability, and decentralisation.

6.1 DeWS: A Decentralised Request-Reply Framework

The request-reply includes two entities: server and clients. The server is responsible for providing digital services to clients. Examples of services include registering a user into a system, handling payment transactions, and sending notifications to supply chain stakeholders when the product reaches a specific warehouse. Typically, a client issues a request to trigger a service on a server. Upon receiving a request, the server processes the request by executing a program that implements the business process. Figure 7.2 (left) shows the architecture of the request-reply framework. Web2 applications follow this architecture for almost every application. However, the problem with this approach is that the server is fully responsible for correctly processing the clients' requests. Typically, the operations of the server are not visible to the clients and the other business stakeholders, increasing the possibility of malpractice.

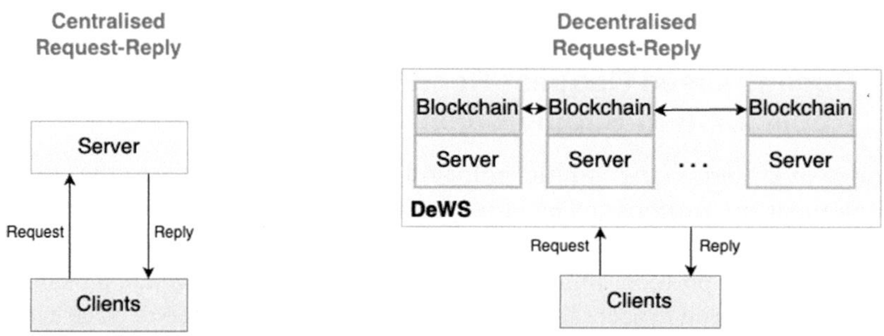

Fig. 7.2 (Left) Centralised Request-Reply with a single point of failure; (Right) DeWS offers a decentralised request-reply framework with blockchain, overcoming a single point of failure.

Figure 7.2 (right) shows the architecture of DeWS. Unlike the centralised request-reply framework, DeWS operate collaboratively in a multi-stakeholder environment. In the case of a supply chain application, each stakeholder, including the farmer, processing facility owner, warehouse operator, and retailer, will have to participate in the deployment of DeWS. Instead of running all services on a single server, DeWS deploys and executes services on a distributed and decentralised infrastructure. Each server is connected to the blockchain, as shown in Figure 7.2 (right). Therefore, when a client issues a request, it is forwarded to the blockchain layer to ensure that each server executes the same request and their responses are validated through a consensus process. If more than two-thirds of the servers produce the same response for a given request, the consensus process concludes successfully, upon which the information is stored on an immutable ledger. This means that all the requests and responses are logged on an immutable ledger, providing transparency and auditability to all the stakeholders. When a client receives a response to its request, they can verify the authenticity and the correctness of the response by checking the immutable ledger. Besides, the clients are assured that the response is generated upon the successful consensus process and storage of data in an immutable ledger.

DeWS guarantees correctness for clients by leveraging the features of blockchain technology. The key benefits of DeWS include the prevention of deliberate computation failures, the generation of incorrect responses, and the dropping of requests. Since the responses are generated upon successful blockchain consensus and storage, malicious servers cannot cheat in DeWS, making it a suitable framework for social-good applications.

7. Whistleblower: A Blockchain-based Machine Learning Framework for Fake News Detection

The widespread adoption of social media applications provides opportunities for malicious individuals to spread fake news. Studies indicate that fake news may have played a critical role in the Brexit referendum and the 2016 U.S. presidential election (Zhou and Zafarani, 2020). Besides, it has been shown that fake news spreads faster than the truth (Vosoughi et al., 2018), further accentuating the social impact. Therefore, fake news can have a huge impact on society, demanding effective solution.

Many practical solutions are being employed. For example, X (formerly known as Twitter) allows users to report fake news by including factual evidence. When users encounter fake news and notice factual evidence countering it, they can ascertain the correctness and form sensible opinions. Such solutions require the involvement of knowledgeable users. Alternatively, machine learning is also employed to detect fake news more effectively. Several machine-learning approaches have been proposed in the literature for fake news detection (Collins et al., 2021; Mridha et al., 2021). Although many of these approaches have proven to be effective, it is unclear whether machine-learning models can operate in unbiased fashions, as studies argue that models tend to discriminate against certain users, regions, etc (Mehrabi et al., 2021). Therefore, it is important to create a system for fake news detection that operates unbiased while involving users in the vetting process. WhistleBlower (Ramachandran et al., 2020) tackles this problem by proposing a democratic and decentralised framework to detect fake news on online social media platforms.

7.1 WhistleBlower: A Decentralised Fake News Detection Platform

Blockchain technology includes features such as consensus protocol, which allows nodes to agree on the system state, smart contracts to run decentralised computations, and the immutable ledger to store records in a tamper-proof manner. These features are suitable for building decentralised applications for social good.

However, it is important to architect the framework carefully to ensure that the end application is truly leveraging blockchain's powerful features. WhistleBlower is an example of a platform that leverages blockchain technology, machine learning algorithms, community members, and incentive schemes to develop an impactful framework for fake news detection.

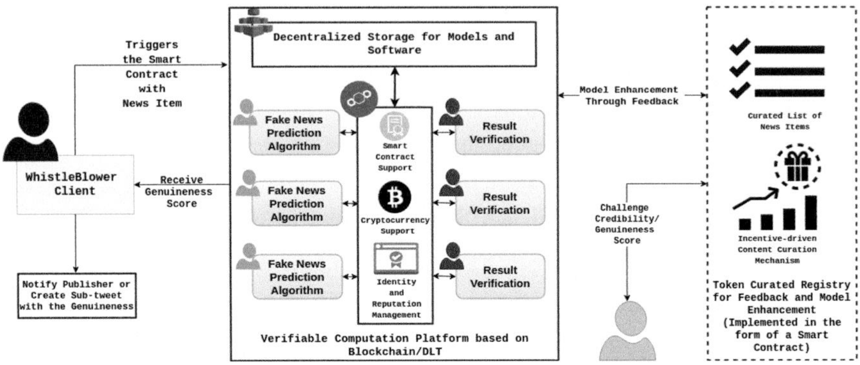

Fig. 7.3 Architecture of WhistleBlower (Ramachandran et al., 2020).

At its core, WhistleBlower is built around a blockchain platform that supports smart contracts, cryptocurrency, and identity and reputation management functionalities. Smart contracts are used to manage the participants and the fake news detection algorithms. Since WhistleBlower involves community members, it is important to incentivise them to increase engagement. Cryptocurrency functionality is used to build tokens to reward participants. Lastly, identity and reputation management functionality are used to keep track of the community participants and their behaviour on the platform.

WhistleBlower employs machine learning to predict whether a given news item is fake or not. The machine learning algorithm comes from the algorithm developers in the community. They can submit their fake news detection algorithms to WhistleBlower, which then employs the best algorithm from the list of community-developed algorithms. By community, WhistleBlower refers to the machine learning researchers and other organisations that are combatting fake news.

Clients typically submit a news item to WhistleBlower to validate its credibility. WhistleBlower platform uses the best community-developed algorithm from its list to predict whether the news item is fake or not. This prediction process is not executed within the blockchain's smart contracts. Instead, the smart contracts choose *N* nodes from a list of computing nodes that are owned and managed by community members. These selected nodes would act as solvers and verifiers. The solvers receive the fake news item and the code for the prediction algorithm. It then performs the prediction by running the computation. Upon completing the computation, the results will include a genuineness score, which will indicate whether the news article is fake or not. However, there is no guarantee that the solver nodes executed the prediction operations correctly. Therefore, WhistleBlower uses verifier nodes to check the computation's correctness. WhistleBlower uses verifiable computation to validate the proof and the results submitted by the prover nodes. For brevity, the details of the verifiable computation are not discussed extensively here and are not included in this chapter, but they can be found in our other publication (Ramachandran et al., 2020).

Upon successful verification of the computation correctness, WhistleBlower returns the genuineness score to the client that initiated the verification process. The client can then report this score to the social media platform through which the fake news is spreading. If the client is not satisfied with the score returned by WhistleBlower, it can challenge the algorithm that was used for fake news prediction. The challenge process is handled by a token-curated registry, which vets the quality of the algorithm by inviting the community members to participate in a review process. The token-curated registry is a content moderation framework built around blockchain technology to assess the quality of content using community members. It rewards the community members for their participation in the moderation process. WhistleBlower relies on the token-curated registry to review the quality of the fake news detection algorithms. Upon completion of the review process, the algorithm is accepted for fake news detection if reviewers are in favour of the algorithm. However, negative ratings from reviewers would result in the algorithm being removed from WhistleBlower. By involving community members in the vetting process, WhistleBlower ensures that biased algorithms are not used for fake news detection.

In summary, WhistleBlower demonstrates how blockchain can be used to build a machine-learning application for fake news detection that is decentralised and transparent. Given the prominence of machine-learning in many real-world applications, a platform-like WhistleBlower can be useful to ensure that unbiased algorithms are used.

8. Reference Architecture

The blockchain-based frameworks discussed in previous sections demonstrated the ability to decentralise applications for social good while supporting legacy communication protocols such as pub-sub and request-reply. Besides, WhistleBlower introduced an architecture to run complex AI and ML algorithms in a decentralised blockchain infrastructure without compromising transparency and accountability. These frameworks collectively provide the following functionalities to develop impactful social-good applications:

Reliable and auditable data exchange and storage: Data is integral to any social-good application operating in a digital space. When data is exchanged between parties, it needs to happen through a reliable and transparent channel. Besides, the data needs to be stored in a database for audit purposes. Trinity and DeWS provide support for data exchange and storage through popular pub-sub and request-reply communication models.

Verifiable computing for arbitrary computations: In addition to the data, computation is critical. In many real-world use cases, the raw data must be processed or combined with multiple data sources to gather business insights. Besides, many applications employ ML and AI models for predictions. These activities require running computations following a reliable and verifiable protocol. WhistleBlower provided an architecture for decentralised and verifiable computation.

Incentives for rewarding good behaviours: Some social-good applications may require support for token mechanisms to reward users, as demonstrated by WhistleBlower.

By combining these functionalities and the key building blocks from Trinity, DeWS, and WhistleBlower, we present a reference architecture in Figure 7.4, targeting social-good applications. To simplify the design process, the decentralised architecture follows different layers. It is a multi-stakeholder framework operating in a collaborative setting, wherein multiple stakeholders contribute resources for computation, communication, consensus, and storage. Figure 7.4 shows a setup with *N* stakeholders. The application functionalities are executed within this layered architecture, wherein each layer is responsible for specific functionalities, as described below:

Data Communication Layer: This layer is responsible for providing *reliable and auditable data exchange.* It builds on Trinity (Ramachandran et al., 2019) and

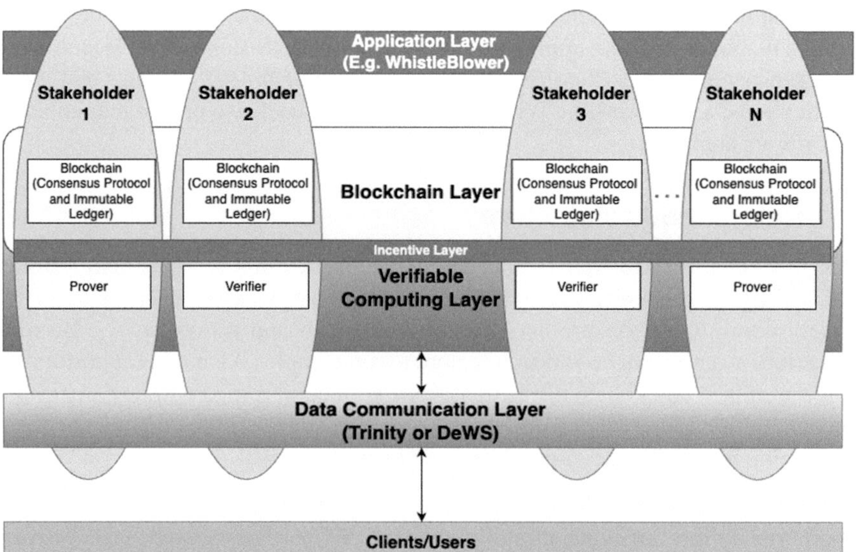

Fig. 7.4 A Reference Architecture for Blockchain-based Social Good Applications. It builds on Trinity, DeWS, and WhistleBlower.

DeWS (Ramachandran et al., 2023). Based on the application requirements, the developers can choose between pub-sub (Trinity) or request-reply (DeWS), making this architecture versatile for real-world applications. For example, applications that only require message exchange among different participants can leverage the Trinity (Ramachandran et al., 2019) framework, while decentralised applications that demand simple computations as part of the data transaction can employ DeWS (Ramachandran et al., 2023).

Verifiable Computing Layer: This layer is responsible for executing computations in a reliable manner. As shown in Figure 7.4, the stakeholders can take the role of a prover or a solver, as discussed in WhistleBlower (Ramachandran et al., 2020). There may be an argument that smart contracts provide mechanisms for verifiable computations, but it has limitations – smart contracts lack the ability to run arbitrary computations because they run in a specialised execution environment such as Ethereum Virtual Machine (EVM). EVM lacks support for machine learning because it is both computationally expensive and requires support for floating point operations. Having a verifiable computing layer outside the blockchain is beneficial because it supports all kinds of computations, unlike smart contracts. Note that WhistleBlower presents an architecture for verifying the correctness of arbitrary computations while respecting the principle of blockchain.

Incentive Layer: Blockchain platforms with support for cryptocurrencies enable application developers to easily integrate tokens to reward users. This layer operates within the blockchain to allow users to transact tokens. Social-good applications often require the continuous involvement and engagement of community members. By rewarding users, social-good applications can maximise the role of community members.

Blockchain Layer: Other layers require the support of the blockchain layer to provide verification guarantees for data and computation while storing all application logs in the immutable ledger for auditability and transparency.

Each of these layers is capable of supporting any digital application that demands transparency, accountability, and auditability. Note that all applications generally include different stakeholders generating data, performing computation on data, making business decisions based on data, and receiving data as alerts. Figure 7.5 shows how the layers are selected and included in a social-good application based on the application requirements, where the blockchain is the only common layer as it is critical for transparency, auditability, and decentralisation. The data communication layer can be beneficial to applications that only require different stakeholders to collectively exchange data through an auditable and decentralised medium. Applications in this category can exclude all the other layers while operating within a blockchain infrastructure. Additionally, depending on whether the existing infrastructure supports a pub-sub or request-reply communication model, further customisations can be made. Recall that Trinity is a pub-sub framework, while DeWS presents a request-reply framework on top of a decentralised infrastructure.

Application Requirements	Layers			
	Data-only	Compute-only	Incentive-only	Decentralisation using Blockchain
Data Transparency and Auditability	✓	X	X	✓
Data and Compute Transparency and Auditability	✓	✓	X	✓
Data and Compute Transparency and Auditability with Incentives	✓	✓	✓	✓
Decentralised Data Marketplace	✓	X	✓	✓
Decentralised Compute Marketplace	X	✓	✓	✓

Fig. 7.5 The table shows how different layers from the reference architecture can be selected based on application requirements.

When applications require computations on the data produced by various stakeholders, the architecture can be extended to include the verifiable computing layer inspired by WhistleBlower. Each stakeholder can collaboratively establish the computation jobs that are essential to process the data. Subsequently, a random set of nodes can be elected to perform the computations. Upon successful completion, some nodes will take the role of a verifier and ensure that the results are correctly computed while leaving proofs and traces on the blockchain for auditing purposes.

Furthermore, the incentive layer can be added to applications that require a rewarding mechanism to encourage crowd participation. The inclusion of this layer in a given social-good application largely depends on the stakeholders and the operational model.

The reference architecture shown in Figure 7.4 can be used to build a wide range of applications. As shown in Figure 7.5, selected blockchain, data, and incentive layers can also develop applications that incentivise stakeholders to share data. This class of application is referred to as data marketplaces (Ramachandran and Krishnamachari, 2023; Ramachandran et al., 2019). Additionally, applications that rely on computations from third parties, such as cloud providers, can leverage blockchain technology to incentivise computation resource providers (Ramachandran et al., 2020). Figure 7.5 depicts this modality, demonstrating the versatility of the proposed reference architecture.

The architecture shown in Figure 7.4 helps the application developers build data-only, compute-only, and incentive-only applications or a combination of those while fulfilling essential requirements such as transparency, auditability, and decentralisation for social good applications. While building blockchain-based social good applications, it is important to consider data governance to ensure that the rights of the data owners are respected (Garcia et al., 2022). Additionally, data quality is critical as the true benefit can come from storing the right data on the blockchain due to the Garbage-In-Garbage-Out problem (Powell et al., 2022).

Conclusion

In conclusion, this chapter provided a comprehensive overview of how blockchain technology can be applied for social good, addressing both technical and practical challenges. It explained how the foundational elements such as public-key cryptography, consensus protocol, immutable ledger, and smart contracts make the blockchain fairer for decentralised applications. More importantly, we highlighted how the blockchain offers transparency, accountability, and decentralisation in applications ranging from supply chain management to misinformation detection.

Subsequently, the chapter also introduced three different decentralised frameworks, targeting data-driven, computation-oriented, and incentive-based applications. Trinity framework introduced how legacy communication protocols such as pub-sub can be integrated with the blockchain to enable decentralised trust in multi-stakeholder applications. Trinity's application to racial profiling was also discussed, emphasising how different law enforcement agencies can collaboratively collect and manage traffic stop data using a decentralised and blockchain-based framework.

Furthermore, DeWS, a decentralised data-sharing framework with request-reply support, was presented to support legacy applications that operate in a traditional Web2 environment. By utilising DeWS, the application data can be managed by a set of diverse stakeholders through a blockchain infrastructure, increasing auditability and transparency. Depending on the application requirements and the infrastructure limitations, the developers can choose between pub-sub and request-reply or a combination of both to build decentralised social-good applications that are data-centric.

For applications that require computation support, WhistleBlower was introduced with a novel verifiable computation to ensure that application data is reliably processed in a decentralised multi-stakeholder application environment. More specifically, the computation jobs are performed by provers, while verifiers ensure integrity and correctness in a decentralised and auditable infrastructure.

Key components like DeWS, Trinity, and Whistleblower, each serving unique roles in data sharing, computation, and incentivisation, are integrated into a reference architecture for blockchain-based applications. This architecture ensures reliable, auditable data exchange, verifiable computing for complex computations, and incentive mechanisms for rewarding positive behaviours. Ultimately, the framework sets a foundation for developing transparent, democratic, and accountable applications, maximising the potential of blockchain technology in enhancing social welfare and ensuring fairness across diverse user groups.

References

Alam, S., Shuaib, M., Khan, W.Z., Garg, S., Kaddoum, G., Hossain, M.S. and Zikria, Y.B. (2021). Blockchain-based initiatives: current state and challenges. *Computer Networks*, *198*, 108395.

Andola, Nitish, Vijay Kumar Yadav, S. Venkatesan, and Shekhar Verma. "Anonymity on blockchain based e-cash protocols—A survey." *Computer Science Review* 40 (2021): 100394.

Antonopoulos, A.M. and Wood, G. (2018). *Mastering ethereum: building smart contracts and dapps*. O'reilly Media.

Auer, R., Frost, J., Gambacorta, L., Monnet, C., Rice, T. and Shin, H.S. (2022). Central bank digital currencies: motives, economic implications, and the research frontier. *Annual review of economics*, *14*(1), 697–21.

Buchman, E. (2016). Tendermint: Byzantine fault tolerance in the age of blockchains (Doctoral dissertation, University of Guelph).

Cao, S., Powell, W., Foth, M., Natanelov, V., Miller, T. and Dulleck, U. (2021). Strengthening consumer trust in beef supply chain traceability with a blockchain-based human-machine reconcile mechanism. *Computers and electronics in agriculture*, *180*, 105886.

Cartier, L.E., Ali, S.H. and Krzemnicki, M.S. (2018). Blockchain, Chain of Custody and Trace Elements: An Overview of Tracking and Traceability Opportunities in the Gem Industry. *Journal of Gemmology*, *36*(3).

Castro, M. and Liskov, B. (2002). Practical byzantine fault tolerance and proactive recovery. *ACM Transactions on Computer Systems (TOCS)*, *20*(4), 398–61.

Collins, B., Hoang, D.T., Nguyen, N.T. and Hwang, D. (2021). Trends in combating fake news on social media—a survey. *Journal of Information and Telecommunication*, *5*(2), 247–66.

Fan, C., Lin, C., Khazaei, H. and Musilek, P. (2022, August). Performance analysis of hyperledger besu in private blockchain. *Proceedings of 2022 IEEE international conference on decentralized applications and infrastructures (DAPPS)* (pp. 64–73). IEEE.

Garcia, R.D., Ramachandran, G. and Ueyama, J. (2022). Exploiting smart contracts in PBFT-based blockchains: A case study in medical prescription system. *Computer Networks*, *211*, 109003.

Garcia, R.D., Ramachandran, G.S., Jurdak, R. and Ueyama, J. (2022). Blockchain-aided and privacy-preserving data governance in multi-stakeholder applications. *IEEE Transactions on Network and Service Management*, *19*(4), 3781–93.

Garcia-Teruel, R.M. (2020). Legal challenges and opportunities of blockchain technology in the real estate sector. *Journal of Property, Planning and Environmental Law*, *12*(2), 129–45.

Garg, Nishant. Apache kafka. Birmingham, UK: Packt Publishing, 2013.

Ge, Lan, Christopher Brewster, Jacco Spek, Anton Smeenk, Jan Top, Frans Van Diepen, Bob Klaase, Conny Graumans, and Marieke de Ruyter de Wildt. Blockchain for agriculture and food: Findings from the pilot study. No. 2017-112. Wageningen Economic Research, 2017.

Hunkeler, U., Truong, H.L. and Stanford-Clark, A. (2008, January). MQTT-S—A publish/subscribe protocol for Wireless Sensor Networks. In 2008 3rd International Conference on Communication Systems Software and Middleware and Workshops (COMSWARE'08) (pp. 791–98). IEEE.

Isaak, J. and Hanna, M.J. (2018). User data privacy: Facebook, Cambridge Analytica, and privacy protection. *Computer*, *51*(8), 56–59.

Jurdak, Raja, Ali Dorri, and Mahinda Vilathgamuwa. "A trusted and privacy-preserving internet of mobile energy." IEEE Communications Magazine 59.6 (2021): 89–95.

Kahya, A., Avyukt, A., Ramachandran, G.S. and Krishnamachari, B. (2021, July). Blockchain-enabled Personalized Incentives for Sustainable Behavior in Smart Cities. *Proceedings of 2021 International Conference on Computer Communications and Networks (ICCCN)* (pp. 1–6). IEEE.

Mehrabi, N., Morstatter, F., Saxena, N., Lerman, K. and Galstyan, A. (2021). A survey on bias and fairness in machine learning. *ACM computing surveys (CSUR)*, *54*(6), 1–35.

Mridha, M.F., Keya, A.J., Hamid, M.A., Monowar, M.M. and Rahman, M. S. (2021). A comprehensive review on fake news detection with deep learning. *IEEE Access*, *9*, 156151–156170.

Nakamoto, S. (2008). Bitcoin: A peer-to-peer electronic cash system. Decentralized business review.

Natanelov, V., Cao, S., Foth, M. and Dulleck, U. (2022). Blockchain smart contracts for supply chain finance: Mapping the innovation potential in Australia-China beef supply chains. *Journal of Industrial Information Integration*, *30*, 100389.

Natanelov, V., Cao, S., Foth, M. and Dulleck, U. (2022). Blockchain smart contracts for supply chain finance: Mapping the innovation potential in Australia-China beef supply chains. *Journal of Industrial Information Integration*, *30*, 100389.

Powell, W., Cao, S., Miller, T., Foth, M., Boyen, X., Earsman, B., ... and Turner-Morris, C. (2021). From premise to practice of social consensus: How to agree on common knowledge in blockchain-enabled supply chains. *Computer Networks*, *200*, 108536.

Powell, W., Foth, M., Cao, S. and Natanelov, V. (2022). Garbage in garbage out: The precarious link between IoT and blockchain in food supply chains. *Journal of Industrial Information Integration*, *25*, 100261.

Ramachandran, G., Nemeth, D., Neville, D., Zhelezov, D., Yalçin, A., Fohrmann, O. and Krishnamachari, B. (2020, November). WhistleBlower: towards a decentralized and open platform for spotting fake news. In 2020 IEEE International Conference on Blockchain (Blockchain) (pp. 154–161). IEEE.

Ramachandran, G.S. and Krishnamachari, B. (2023). Blockchain for Data Sharing. Blockchains: A Handbook on Fundamentals, Platforms and Applications, 601–25.

Ramachandran, G.S., Contreras, S.L., Krishnamachari, B., Kozat, U.C. and Ye, Y. (2019). {Publish-Pay-Subscribe} Protocol for Payment-driven Edge Computing. In 2nd USENIX Workshop on Hot Topics in Edge Computing (HotEdge 19).

Ramachandran, G.S., Malik, S., Pal, S., Dorri, A., Dedeoglu, V., Kanhere, S. and Jurdak, R. (2022). Blockchain in supply chain: Opportunities and design considerations. In *Handbook on Blockchain* (pp. 541–576). Cham: Springer International Publishing.

Ramachandran, G.S., Radhakrishnan, R. and Krishnamachari, B. (2018, September). Towards a decentralized data marketplace for smart cities. *Proceedings of 2018 IEEE International Smart Cities Conference (ISC2)* (pp. 1–8). IEEE.

Ramachandran, G.S., Wright, K.L., Zheng, L., Navaney, P., Naveed, M., Krishnamachari, B. and Dhaliwal, J. (2019, May). Trinity: A byzantine fault-tolerant distributed publish-subscribe system with immutable blockchain-based persistence. In 2019 IEEE International Conference on Blockchain and Cryptocurrency (ICBC) (pp. 227–35). IEEE.

Ramachandran, Gowri, and Raja Jurdak. "DeWS: A Decentralized and Byzantine Fault-tolerant Web Services." In The Proceedings of the 5th IEEE International Conference on Blockchain and Cryptocurrency, Dubai, UAE, May, 2023.

Saleh, F. (2021). Blockchain without waste: Proof-of-stake. *The Review of financial studies*, *34*(3), 1156–90.

Tang, S., Wang, Z., Dong, J. and Ma, Y. (2022). Blockchain-enabled social security services using smart contracts. *IEEE Access*, *10*, 73857–73870.

Tharatipyakul, A., Pongnumkul, S., Riansumrit, N., Kingchan, S. and Pongnumkul, S. (2022). Blockchain-based Traceability System from the Users' Perspective: A Case Study of Thai Coffee Supply Chain. IEEE Access, 10, 98783-98802.

(Stackpole, 2022) Stackpole, Thomas. "What is web3." Harvard Business Review 10 (2022).

Trotter, L., Harding, M., Shaw, P., Davies, N., Elsden, C., Speed, C., ... and Hallwright, J. (2020, December). Smart donations: Event-driven conditional donations using smart contracts on the blockchain. *Proceedings of the 32nd Australian Conference on Human-Computer Interaction* (pp. 546–57).

Vosoughi, Soroush, Deb Roy, and Sinan Aral. "The spread of true and false news online." Science 359, no. 6380 (2018): 1146–51.

Zhou, Xinyi, and Reza Zafarani. "A survey of fake news: Fundamental theories, detection methods, and opportunities." ACM Computing Surveys (CSUR) 53, no. 5 (2020): 1–40.

8 Non-fungibility as a Conceptual Resource for Design

Chris Speed[1*], Chris Elsden[2] and Peter Tilley[2]

1. Introduction

Despite close to two decades of cryptocurrency and blockchain developments, fungibility remains a defining characteristic of almost all fiat currencies in circulation today. The term 'currency' is derived from the latin word "currere," which means "to run" or "to flow", and for economic value to run or flow, it requires as few barriers as possible, in other words, we require it to be fungible. In economic terms, fungibility is a key characteristic that allows value to flow freely within a currency system. The principle of fungibility refers to an interchangeability between units of a particular asset or commodity. In the context of a government backed fiat currency, this means that a monetary unit is identical and can be exchanged with another unit of the same denomination. For example, a one-pound coin is fungible with another one-pound coin, because they hold the same monetary value. In fact, a one-pound coin can be exchanged for any number of smaller coins as long as they add up to the same amount as a one-pound coin.

Traditional economists will assert that fungibility is a vital characteristic of a currency that supports free flowing, seamless transactions, without the need for individuals to assess historical, social, material or environmental values of a unit of currency. This is a representation of value 'without strings attached'—or at least very few, other than proof that it is issued by a national bank, that it is not a forgery or counterfeit, and that it has information on it describing the units of value that it represents. Now common in many forms from physical cash, or as a digital transaction performed through the tap of a smartphone, the fungible nature of currency streamlines economic transactions and provides a commonly accepted medium of exchange. Such characteristics enable multiple forms of value to be developed within society extending beyond the economic, including trust, speed of exchange, and standardised forms of record keeping leading to debt and credit.

Economists have long maintained that an interdependency between fungibility and currency provides a fundamental architecture for a standardised and universally accepted medium of exchange, essential for functioning modern economies. In

[1] RMIT University, Melbourne, Australia.
[2] Institute for Design Informatics, The University of Edinburgh, UK.
Email: chris.elsden@ed.ac.uk[2]; ptilley395@hotmail.com[3]
* Corresponding author: chris.speed@rmit.edu.au

contrast, social and cultural scholars introduce further dimensions to both the practical and historical concept of money. For example, relational sociologists, following in the footsteps of Zelizer's essential work on the 'social meaning of money' (1989), emphasise firstly that fungibility is an idealistic concept of money.

In reality, even in high-frequency stock market trades, there are numerous frictions at play and strings attached, such as the latency involved between the execution of a trade and its confirmation, the risk that the other party in a trade might default on their obligations, and the presence of other high-frequency traders competing for similar opportunities that can reduce profitability (Mackenzie, 2022). Likewise, despite their interchangeability, no two £10 notes are truly the same, and in addition to bank authorities adding unique serial numbers and in some countries a date, Zelizer shows numerous subtle social practices through which money is 'earmarked' and differentiated. As such, rather than a discrete binary of 'fungible' and 'non-fungible,' the authors of Money Talks (Bandelj, Wherry and Zelizer, 2017) describe fungibility as a variable quality of money and exchange.

O'Dwyer recovers the concept of money as memory and its shifting potential as the digital economy introduces new ways to 'cache' or store where and who money has interacted with (O'Dwyer, 2019). Through an eloquent reminder of the etymological roots of the term money—from Moneta, a translation of the Greek Mnemosyne who was the goddess of memory—O'Dwyer cites Hart in reframing money as an 'instrument of collective memory' (2001, p. 234). By charting four 'orders' of economic memory from the recording of assets, to contracts, to accounts, and proxies, O'Dwyer establishes the historical 'mnemonic infrastructures' of money. But her final move explains the implications of these architectures as they become actionable through data-driven technologies. Money as memory, previously siloed and in the custodianship of merchants with limited agency, is now in the hands of platform economies in the form of data; "Platforms are the new memory banks" (2019, p.149).

This chapter begins with an introduction, establishing the significance of the concept of fungibility in economic systems and its historical context. The section explores various perspectives from traditional economists to social and cultural scholars, highlighting the complexities and limitations of fungibility. A methodology section outlines the design and implementation of a virtual workspace to explore non-fungibility as a conceptual resource, which in turn leads into the case studies section, which presents three design research projects: Tales of Things, KASH Cups, and A Token Gesture. Each case study illustrates how non-fungibility can create multiple forms of value through innovative design practices.

The discussion and contributions section reflects on the implications of adding social and ecological values to currency, challenging the concept of fungibility, and proposing non-fungibility as a valuable resource for design. Finally, the chapter concludes by suggesting future directions for research and practice, emphasising the potential for non-fungible design to support sustainable and regenerative societies.

2. The Climate Crisis as a Significant String

As platforms for memory, large scale digital economies have banks full of not only economic data but also deep insights into our personal, social and political interests. Social, personal, environmental attributes combine to form complex data shadows that surveil our actions and attempt to anticipate our intentions and desires (Zuboff, 2019). Such contemporary conditions challenge the very possibility of fungibility or 'value without strings attached.' Whilst a modern economy relies on fungibility to support the interchangeability of economic units, any consideration of social or environmental values associated with a transaction could represent significant 'strings' that impede its ability to flow.

The UN's Sustainable Development Goals (SDGs), which offer interlinked objectives designed to move society towards social, environmental and economic well-being, demonstrate an increasing interest in understanding the ethical and ecological provenance and consequences of economic activities. If transactions are linked to socially and environmentally harmful practices, it calls into question the very concept of fungibility—in particular, if it slows down the flow of value across a currency. Is it possible that individuals and institutions may start to consider the social and environmental history of money, making distinctions between currency units based on their ecological impact?

As further pressures on financial services organisations grow to introduce measures to ensure the Environmental and Social Governance (ESGs) of the flow of economic value, so called 'fungible currencies' could become increasingly untenable, as not all units of value would be perceived as equal if they carry different environmental histories. Evidence of this trend is demonstrated in the five-year review of SWIFT (Society for Worldwide Interbank Financial Telecommunications) and its own progress towards the full adoption of ESG principles across the finance sector. In many ways SWIFT provides the data 'strings' that are associated with financial transactions to ensure accuracy and efficacy. Evidently, they are increasingly augmenting not only personal and economic information with a transaction, but environmental and social data:

"I think data (alongside qualitative assessments) will drive ESG ratings. I would say that within ten years, we'll have imperfect ESG data for every service or good we purchase. We are moving toward the democratisation of ESG," Stephanie Gripne (SWIFT, 2023).

As designers, our interest lies in how such conditions may lead to the development of parallel forms of currency that align more closely with the social and ecological values of citizens, challenging the interchangeability that is essential for fungibility. The intersection of the climate crisis and our reliance on fungible 'no strings attached' forms of economic value raises important questions about the future of monetary systems and the need for design innovations that align better with social and environmental values.

At a time when the organisation of society operates in a condition of late capitalism, exacerbated by the rise of platform economies, our paper builds on the work of others that are repositioning design toward addressing global ecological challenges. In their 2017 paper, *Light et al.* set out the tremendous challenges that humanity, in an expanded sense, faces, that encompasses both our co-existence across ecosystems, and our kindness come to the fore (Light et al., 2017). Careful to challenge design's heritage, the work asks us to 'abandon' some of the classic interaction design ideas including propositions that 'users are us', as fallible beings, that 'ease is not serving us' and that instant gratification should be avoided in place of a long term view for interaction, and finally 'Human-Computer Interaction (HCI) is the wrong term' that underlines the redundancy of a title in a time when we acknowledge the entanglement with complex social, technical and more-than-human ecologies (ibid). In their special edition entitled 'Repositioning CoDesign in the age of platform capitalism: from sharing to caring', Avram et al. (2019) curate a series of papers that consider how design interfaces with complex digital economies that have coopted the term 'sharing' to place more emphasis upon the affect of those caught up in within them. Of particular relevance to the work on money, Huttunen and Joutsenvirta (2019) explore the role that complementary currencies (including cryptocurrencies) offer a means of improving economic resilience and social injustice. By engaging directly through data-driven technologies, case studies demonstrate how the underlying architecture of complementary currencies lend themselves to "paying more attention to the modes of collaboration, systemic factors and the social particularities of each project and their environment" (ibid) and in turn point toward sustainable futures.

3. Non-Fungible

As we consider adding social and ecological values to our money through association with claims of authenticity, attribution and provenance, we can accept a shift away from pure fungibility towards less fungible or non-fungible currencies and tokens. Non-fungibility arises when each unit of currency holds unique characteristics (or strings attached) that distinguish it from others, instilling it with particular conditions and making it less interchangeable on a one-to-one basis.

The information a particular unit of currency may be associated with could include the social integrity of the organisations that provide the goods and services that someone is using the money to purchase. Or, it may chart the environmental credibility of goods as they move across a supply chain from source, through manufacture and production, and towards purchase.

The history of each transaction that is likely to be stored digitally and become accessible to a consumer could make units of currency extremely non-fungible, as one unit may no longer be considered equivalent to another due to divergent social and ecological attributes. Previous work of authors, prior to the use of blockchain technology, addressed the range of technologies that have been used to add personal data to physical artefacts (de Jode et al., 2011), and the extent to which these early digital economic practices generated an 'internet of old things' in contrast to the emergence of the 'internet of things' platforms that prioritizes connectivity between devices, and the exchange of quantitative data.

Whilst such 'strings' to a currency sound extremely challenging for the flow of value within an open market, we are interested in the design opportunities and consequences that this presents, as the transparency and accountability of economic transactions begins to challenge and change consumer behaviour. Such a shift could support the transition towards values-based economies in which citizens and organisations prioritise social and environmental values over purely economic. An opportunity that renders the development of non-fungible or less fungible forms of currency, is a unique opportunity for designers to intervene not only in the established design of products and services, but in the currencies and systems of value exchange that both pay for and fund them. However, steps toward the use of non-fungible forms of currency face the challenge of deciding what values to attach to data. If personal information becomes associated with currency, then very quickly the metaphorical 'strings' translate in to forms of tracking which in turn exposes particular members of the community to pernicious forms of surveillance, such as those receiving money through government social security programs (Beilefeld, 2018, 2021). Within 'democratic' economies the existing deployment of non-fungible currencies may be restricted according to particular conditions and communities, whilst a more totalizing representation of value as data is manifest through China's social credit system which, on one hand, attempts to develop trust to combat corruption, but is also designed to promote state-sanctioned moral values – a challenging proposition that could lead to state control (Gruin, 2021; Jia, 2020).

4. Here Comes the Blockchain

To associate social and environmental values with a currency, some form of digital platform is likely required to hold the values in datasets and allow them to be recovered at points of potential transaction. The emergence of blockchains has demonstrated suitability for such a purpose, and although the authors remain open minded to many parallel and historical technologies, it is worth noting the technical characteristics of blockchain technology. Blockchain comprises a decentralised ledger, cryptographically secured, able to record and store information about the nature of transactions associated with a particular unit of currency. Blockchain also makes this information accessible and verifiable by users of the currency within the network (Foth, 2017; Elsden et al., 2018).

Fungibility for cryptocurrencies such as Bitcoin, is represented through their interchangeability and divisibility, as each unit is identical to others and can be divided into smaller, equal components. Alongside their security, the fungibility of cryptocurrencies contributed to their significant growth following the 2008 banking crash. In contrast, non- fungible tokens (NFTs) represent a unique class of digital assets that are indivisible and distinguishable from one another. Each NFT is distinct, contributing to its scarcity and value. This uniqueness has proven particularly valuable across digital art, collectibles, and gaming, where NFTs are used to represent ownership and provenance for digital assets, e.g. 'Faircoin' or other social / community currencies.

Of course, the use of blockchain and its creation of a non-fungible currency leads us quickly to Non-Fungible Tokens and the hype cycle associated with the use of the technology to record unique associations with editions of artwork (Valeonti et al., 2021). However, in an attempt to prevent the most recent boom and bust culture and economy of NFTs in eclipsing our argument and design practice, we return to three design research projects that considered the association of social and environmental values (through the use of data technologies) with the functional and experiential value of a product or service.

5. Conceptualising NFTs as 'Keyrings'

In light (or indeed, shadow) of the hype cycle around NFTs as speculative assets, we have sought alternative ways to discuss, think about and design non-fungible tokens with various participants and stakeholders. One of these is to consider NFTs as a kind of special 'keyring' – a platform, with a series of mechanisms, that allows the holder to undertake specific actions, such as, gaining access, storing documents, proving identity or holding personal mementos. One might hold multiple different keyrings, for different roles and needs, or have many interconnected keyrings for related purposes.

The metaphor helps us consider NFTs as a platform to assign specific rights and actions. It highlights the primacy of physical possession (i.e token holding) as a claim of ownership and means of sharing rights. The ubiquitous Airbnb lockbox shows how leaving your keys with a stranger allows them (at least temporary) access and rights to your home. Just as any stranger can independently prove a kind of ownership by demonstrating control of a wallet associated with a specific NFT.

The way in which we tend to add to and cluster multiple tools on a single keyring, reflects the potentially multiple and extendible applications of an NFT. Finally, it helps us recognise the extent to which NFTs, produced by, and dependent upon interactions with smart contracts[1], are highly fixed and set mechanisms, much like a key and lock (Natanelov et al., 2022). Indeed, an intriguing quality of NFTs is that the nature of the 'key' is public, and therefore anyone can design a new box, door or house that that key can open. For example, though an NFT may be issued for one purpose (e.g. to prove ownership of a digital artwork), it can subsequently be used as a membership scheme, or basis for a voting system.

Taking this metaphor beyond a concept, we worked with a graphic illustrator, Peter Tilley, to produce a template and canvas for 'designing a keyring,' where each loop or key provides particular rights or services, such as access, storage or identity.

Designed originally for use online, using the whiteboard tool Miro, the canvas allows users to dynamically mix and match various 'keys' and attachments, linked on to the rim of an illustrative 'token' at the centre of the keyring (Fig. 8.1).

---

[1] A smart contract is a self-executing program or script that automates the actions required in a block-chain transaction. Once completed, the transactions are trackable and irreversible (Nissen et al., 2018).

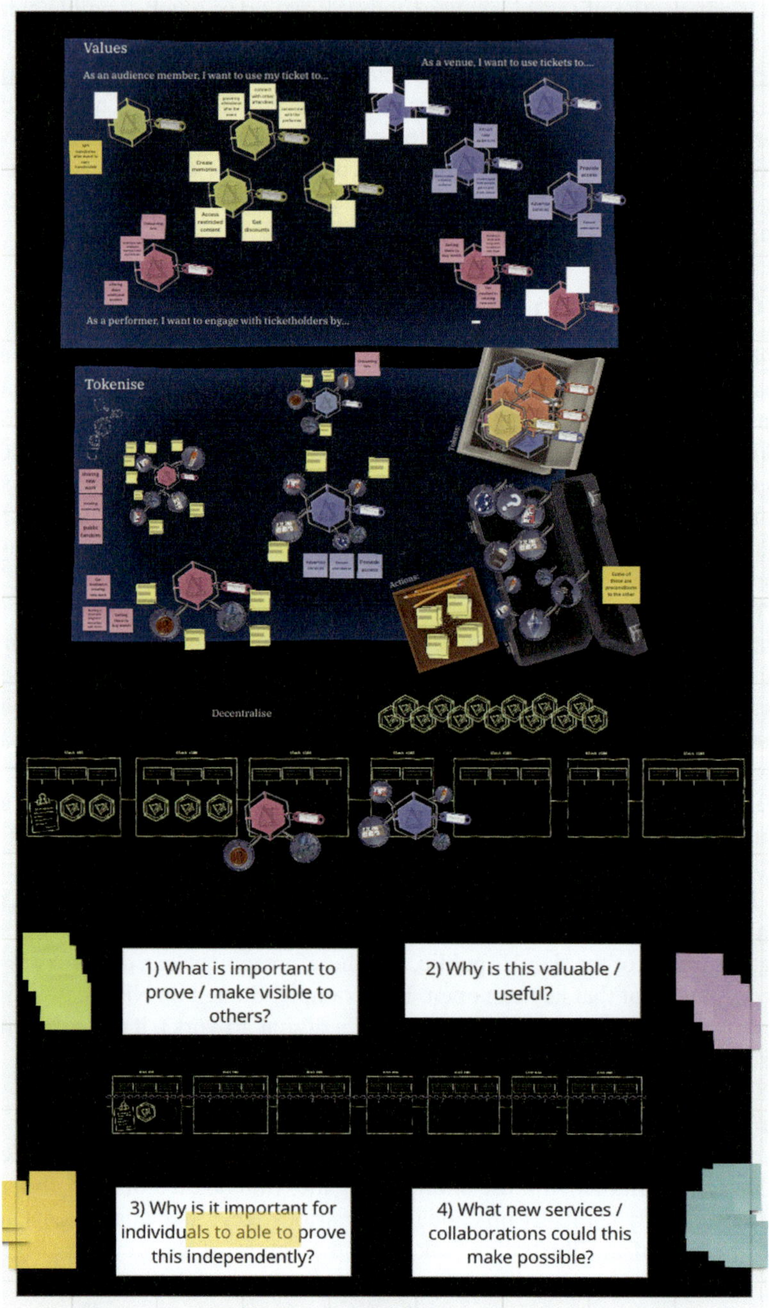

Fig. 8.1 Miro board used to design non-fungible services by linking conditions (values) to a keyring. Image by authors.

Attachments are premised on a series of basic actions or mechanisms that could be enabled through possession of the token – such as 'access' to a certain place or object, voting rights, registration of some important data, proof of identity, or simply a status symbol. The initial attachments are intended to be expanded and are offered only as a starting point to help people think through possible relations and rights that a tokenholder could be granted and claim.

When using the canvas in a workshop context, we have prefaced the building of keyrings with a more rudimentary exercise to discover the 'values' at play in a given context. For example, on one occasion we asked participants to consider a retail scenario where upon buying from a particular shop, one's receipt was also a form of NFT, which could enable the holder to access various other services. A merchant may wish to use the receipt to build loyalty, encourage referrals, or prove the environmental credentials of their goods; the customer may value a sense of connection or exclusivity, future economic benefits and rewards, or the ability to demonstrate ethical shopping habits. Upon this basis, we could then ask: how could we design an NFT keyring to mediate some of these values? What are the range of potential activities and rights that could be granted and claimed?

What this really becomes is an exercise in exploring how specific values can be associated with a particular transaction or exchange – strings attached – and hence relations and value(s) exchange that this makes possible. Beyond dreaming up schemes for new NFTs, we propose this methodology as a means to explore and develop 'non-fungible design services.' By this we mean a facilitatory toolkit that supports the designer/s in considering how a product or service can be considered to support the co-creation of multiple forms of value.

Each proposition becomes 'non-fungible' because of the interdependence that the new service has within an existing system. Rather than considering a design artefact or service as isolated with no responsibilities, no strings attached – or fungible, instead the designer is encouraged to foresee values becoming realised through reciprocal relationships with other services and agents ranging from data sets, artefacts, technologies, people and communities. In this way, non-fungibility becomes a resource for design, whereby imposing constraints and strings on the way we transact, new relations and meaning can be created. In the context of ESGs, 'Non-fungibility as a resource for design' supplants the ambiguity as a resource for design proposed by Gaver et al. (2003).

Finally, the workshop considered the necessity and implications of decentralisation for these tokens. In the retail context, there are already many customer reward and membership programmes which endeavour to mediate the kinds of values described above. However, crucially, decentralisation can produce a more level playing field, where individuals have a greater stake in the data produced about their transactions and can port this 'data' (reified and verifiable through reference to a distributed ledger) somewhere else. It can potentially be read publicly, engaged with by other merchants or competitors, accessed and shared independently and indefinitely. There are of course profound challenges and implications for this kind of decentralisation, but it prompts us to reflect on how the control and centralisation of data connected to our transactions mediates the values that these transactions can carry and represent.

6. Three Projects that Associate Values with Value Transactions

The association of data with social experiences to challenge and disrupt the value of designed products and services has been a central mission of the Institute for Design Informatics at the University of Edinburgh in the UK from 2016 to the present day (Murray-Rust et al., 2023). With experiences ranging from performing fake public marriages that inscribe a contract between two people onto the bitcoin blockchain, to prototyping hair dryers that reduce the levels of control that a user has in order to manage energy use, each system entangles human intentions within technical, social and environmental intra-actions (Barad, 2003). To offer the reader examples of how values become created and associated through the keyring toolkit, we offer three case studies that are placed within the non-fungible design framework.

Project 1: Tales of Things (National Museum of Scotland, 2011)
Tales of Things (de Jode et al., 2011) was an interactive web service that allowed digital media to be associated with physical artefacts. Funded through a UK research grant between 2009 and 2013, the mobile and online platform enabled individuals to generate QR codes that could contain links to text, audio and video material that were subsequently attached to a physical object of personal or social value. The project became popular in digital heritage and museum communities as it allowed curators to add digital media to artefacts that were otherwise kept behind glass and had limited space for interpretative material.

Whilst museums used the technology to enhance the depth of material for a unique and rare object, the National Museums of Scotland were interested in not just disseminating information, but in gathering stories. Stories associated with mundane and ubiquitous objects that were common to the 'living memory' of growing up in Scotland. Objects and themes varied from Singer sewing machines that were produced in their millions from a factory in the west of Scotland, to the social phenomena of 'dance halls' where many young people met during the 1950s and 60s.

In response to the opportunity, the authors introduced 'ghost objects' (Figure 8.3) as a means of eliciting stories from the public and associating them with an authentic and unique item that is part of the museum's collection. At a series of public events, a selection of ghost objects was offered to visitors to encourage them to recall stories associated with the original artefacts and Scottish culture. For example, a ghost shoe became a device for eliciting stories about dance halls. By peeling a unique barcode from a sheet of stickers, the public could scan it, and directly associate it with a text, video or audio version of their story. Each contributor was then invited to stick the QR code directly on to the ghost object

Fig. 8.2 A ghost shoe. A ubiquitous high heeled shoe that allowed visitors to the Museum of Scotland to associate their memories of Scottish dance halls with a pair of authentic shoes from the 1950s that is displayed at the museum. Image by authors.

that they were referring to, and in turn their digital media would appear after the primary story associated with the original artefact that sat behind glass in the museum.

In many ways the ghost object became the fungible instantiation of the public memory of high heeled shoes and their association with dance halls, through which personal stories could be told. Whilst the original pair of shoes from the 1950s was held behind glass remains the rare non-fungible version, the assemblage of technologies allowed multiple forms of value to be created:

1. Curators could control the primary story associated with an original artefact – the NFT.
2. Ghost objects allowed secondary stories from visitors to the museum to be publicly associated with the original object (NFT), in different forms – text, video and audio (Speed, 2012).
3. In some cases, the addition of digital media to an object added economic value to an original item. The gathering of public stories contributes to the provenance of items.
4. The ghost objects opened a channel for the public to contribute to the social historical value of artefacts in the museum's collection. Not only an interesting social heritage exercise, but also in demonstrating the museum's continuous commitment to community engagement.
5. The Tales of Things web technology at the time allowed the dynamic creation of QR codes that could be associated with any digital material, and developed a large, publicly accessible online corpus of social memory.

The keyring canvas eloquently helps to unpack, and in turn visualise the values that are created and held together across the assemblage of material and digital elements (Figure 8.3).

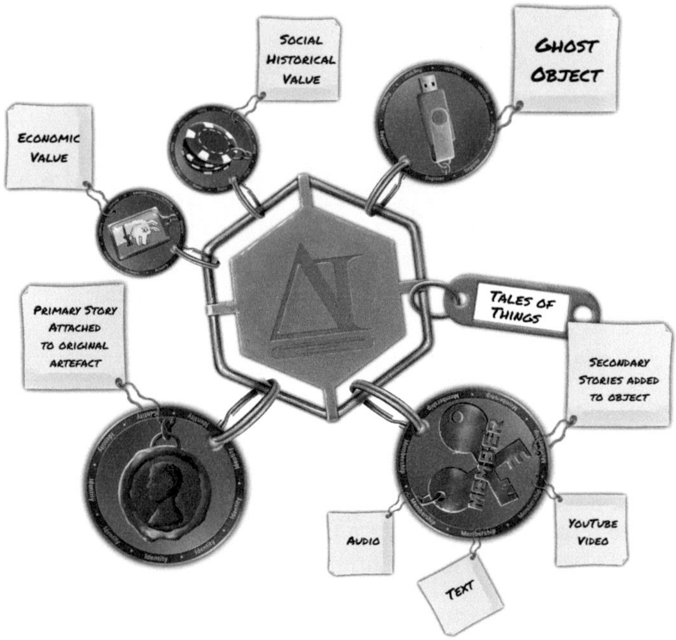

Fig. 8.3 The Tales of Things and ghost object project mapped out as a non-fungible design service. Image by the authors.

Project 2: KASH Cups (Dutch Design Week 2016)

KASH Cups (Murray-Rust et al., 2023) was an interactive system of coffee cups, which aimed to reveal new opportunities for the design of value systems. Use value, economic value and social value, are often disguised in the habitual processes of using money as a representation of value. Augmented with NFC technology, the KASH Cups helped to surface different forms of value as the 'owner' of each cup encountered different people including the barista within the social setting of a conference. The material artefacts (cups) and their programmable functionality challenged the social relationships between people who took part in the event by translating their otherwise informal interactions into transactions across which particular values became stabilised.

Each cup had an NFC tag in its base that provided it with a unique identity within a database. The database, which ran locally on a PC, was able to store two simple conditions: whether the cup had credit to receive a free coffee from the barista, or not. Visitors to the Things2Things event, that ran parallel to Dutch Design Week (Giaccardi et al., 2016), were able to pick up a coffee cup, but in order to receive credit for a coffee had to find someone else who wanted a coffee and together they would place their cups next to each other on a smart saucer to earn credits for coffee (Fig. 8.4). Upon meeting the barista, each delegate placed their KASH cup on a further smart saucer that would detect whether each cup had credit in the database, and if it did, he would proceed to make a coffee of their choice.

Fig. 8.4 Two delegates to the conference working together to add credit to their KASH cups. Photograph by the authors.

This intervention into the social conventions of a conference complicates the traditional value constellation surrounding free coffee, by adding more values to the traditional service. The 'adding of more values' can be mapped onto the NFT Keyring Miro Board to better reveal how the intervention co-creates value in different ways.

Each KASH cup has transactional ties akin to smart contracts that supports different forms of value creation:

1. Free coffee, a mandatory requirement of all good social and especially conference gatherings.
2. The KASH cups were designed and produced by the internationally recognised ceramicist Katy West. The cups were part of a limited series and became valuable as cultural artefacts in their own right.
3. Choosing to interact with a stranger for the mutual benefit of acquiring a free coffee made participants 'friends.' The highly convivial act became an ice breaker to broker new conversations and relationships.
4. A popup coffee service with an experienced barista was hired to make the coffee. Dedicated to the craft of making perfect coffees, the KASH cup service promised great, handmade coffee.
5. A simple SQL database that associated NFC chips in the base of the cups with the smart saucers enabled each KASH cup to become a simple wallet that stored the credit and debit for each delegate.

Unlike encountering any normal empty coffee cup at a conference, which could be understood to be entirely exchangeable and therefore fungible until it is filled with a particular persons preferred coffee, the KASH cups can be considered

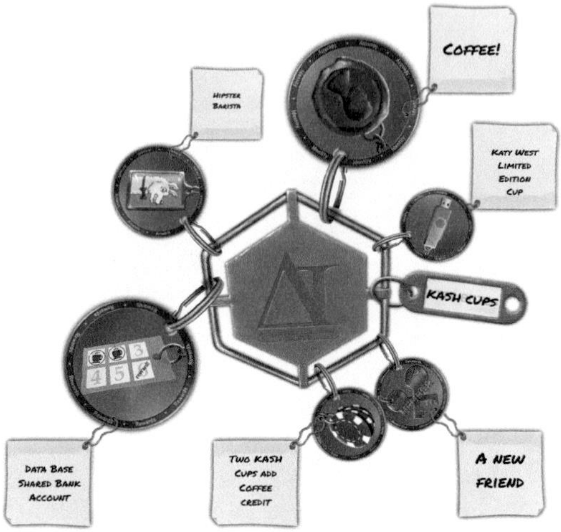

Fig. 8.5 The KASH Cups coffee service mapped out as a non-fungible design service. Image by the authors.

a material instantiation of an NFT. But it is the system of associations with other points of value creation, which represent 'smart contracts,' that offer a non-fungible design service. The NFT keyring canvas enabled us to map the role that the KASH Cup system had in creating and sustaining multiple forms of value (Figure 8.5).

Project 3: A Token Gesture (Inspace City Screens, 2022)

A Token Gesture[2] (Elsden et al., 2024) was an exhibition and research project to introduce, explore and critique new public interactions and ownership of digital art via NFTs. Spanning the boom and bust of NFT projects in the digital art market, this intervention sought to directly engage the public in Edinburgh in the co-creation of digital artworks, and the experience of NFT ownership. The experience enabled passers-by at the Inspace gallery to:

1. Generate unique pieces of digital art through a street-level, walk-up interaction.
2. Register their artwork for display in an exhibition via the Inspace City Screen.
3. Mint, claim and own a non-transferable NFT, representing their piece of art, that will allow them to control how their piece is displayed.

Working with two Scotland-based generative artists – Sasha Belitskaja and Cameron "Gingey", we created a system where anyone can generate a unique piece of digital art simply by presenting a colour to a fixed camera. Participants were then carefully guided through setting up a crypto wallet to claim and mint an NFT representing their piece.

Crucially, this NFT – a unique, digital token – was non-transferable. It could not be exchanged or sold. However, it served to register the artwork immutably,

2 https://nft.inspace.ed.ac.uk/about

evidence an individual's contribution, and to allow them to control when the artwork is displayed on the Inspace City Screen projectors (part of the Institute for Design Informatics, Edinburgh).

More than 200 people engaged with the exhibition, producing more than 1,788 artworks, of which 79 were ultimately minted as an NFT and displayed in the exhibition. In subsequent surveys and interviews, we found that by removing any financial or economic value to the NFT, participants derived other kinds of value from their participation and holding a unique, non-fungible token associated with their artwork (Elsden et al., 2024). Primarily, the NFT produced, functioned as a bespoke 'proof-of-participation,' in a unique project that could subsequently be shared publicly and privately, and demonstrated engagement in a particular community and sub-culture. Indeed, being literally non-transferable, and bound to a single cryptocurrency wallet, this is a rather extreme example of non-fungibility – but this is what allows the token holder to 'prove' that they were there.

Fig. 8.6 The Token Gesture exhibition and 'NFT cash point' where passers-by could mint their own artwork. Inspace City Screen, Institute for Design Informatics, Edinburgh. Photograph by the authors.

The exhibition concluded after four weeks, during which time we organised some brief, public, online events to reflect on and discuss aspects of the exhibition and research project. Prior to launching, and during the exhibition, as a team we considered how we might more explicitly engage those who had taken part in the exhibition and co-create further value with tokenholders. Notably, while all artworks, tokens and their corresponding wallets were visible on a public distributed ledger (Tezos), each tokenholder was effectively anonymous and unknown to each other.

Unfortunately, time and resource constraints did not allow us to pursue these plans. However, here, we wish to speculatively turn to the keyrings metaphor to consider how we might have designed further relations, interactions and services, based on the ownership, and extreme non-fungibility of a Token Gesture NFT.

Once the participant has rights over a particular artwork we envisioned providing them with two forms of value creation:

Firstly, a set of actions around controlling artwork and exhibition:

1. E.g. include or remove the artwork.
2. Managing the screens displaying the artwork (effectively curating the exhibition itself).
3. Removing or limiting subsequent uses of the artwork or the colour input used to create it (since the colour code is included in NFT metadata, this can be a seed or data point for further interaction).
4. Offering voting rights in relation to current or future exhibition content, for example, which artist should be commissioned to produce new work for the exhibition.
5. The ability to subsequently modify or 'mutate' works to produce new artworks.

More broadly, we considered possible value of tokenholders being a membership community of shared interest. Benefits could include:

1. Being airdropped / sent new artworks.
2. Invitations to access exclusive events / activities.
3. Participation in a members-only discussion forum.
4. A badge that provides recognition / community status.
5. A basis to crowdfund or support future artistic work.

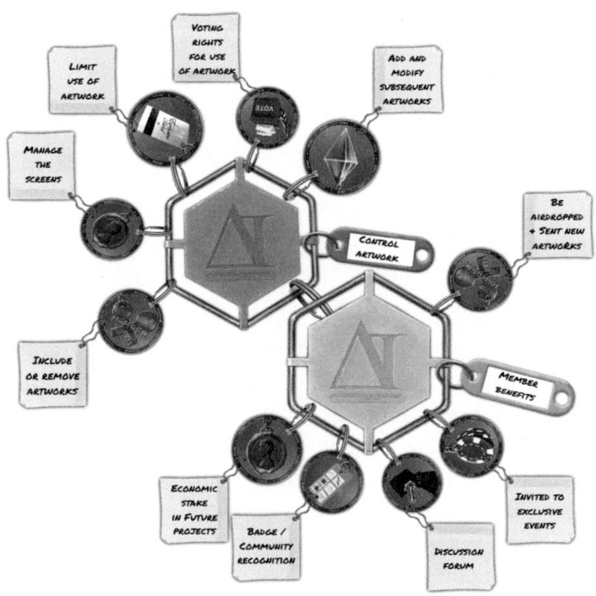

Fig. 8.7 The representation of different forms of value mapped across the two different values (control and community) of the Token Gesture experience. Image by the authors.

These 'values' are represented on the Miro canvas through two interlocked keyrings that provide control over an individual artwork, and benefits of being part of the community (could be a particular collection of artworks, or simply associated with a single artist). Actions associated with each of the two keyrings provide insight into how value is generated across different contracts.

Conclusions

As the collections of Cryptopunks and Bored Apes between 2017 and 2021 became co-opted in the wave of NFTs, they helped to define a particular class of blockchain activity that supported a 'proof' between artistic creators and digital media assets. The definition and use of NFTs recovered the opportunity to ask questions about the spectrum between fungibility and non-fungibility. This chapter has delved a little deeper into a definition of fungibility as a conceptual starting point for currency, if we consider money to be a free-flowing asset. By introducing the implication of complicating a currency by adding data and memory (or values) we suggest that it adds friction to its ability to flow, and in turn it becomes less fungibile and moves towards non-fungibility.

Whilst friction may traditionally be considered an impairment to a currency, in the context of economic systems that should demonstrate increased Environmental and Social Governance (ESG), the data that will become associated with less carbon-intensive transactions will inevitably increase the value of some products and services. In this way the strings attached to a transaction represent positive friction, offering demonstrable application of ESG. The opportunity then arises for designers to initiate any engagement with a partner, client and community towards the co-design of a product or service in such a way that it purposely sets out to design with strings attached – towards non-fungibility. With this in mind, the article introduces the NFT keyring canvas as a means of designing toward or reflecting upon the co-creation of value between existing products and services within a given context. It provides a starting point to consider non-fungibility as a resource for design.

By adding strings to the introduction of a new 'thing,' artefacts and services no longer become the centre of attention, instead they become entangled in the co-creation of value within complex social and environmental ecosystems. Of course, this was always the case, and it was probably the conceit of all designers to turn a blind eye to the implications that their contributions to society had, if we consider them within wider value constellations (Speed and Maxwell, 2015).

The adoption of decentralised financial technologies has not only challenged our conceptual models for what money is, what it can do, and its limitations, but the theoretical movements involved with designing with data toward new forms of value transactions, changes how we consider both products and services.

The limitations of the work remain around the sampling of only three design research projects, and the team look forward to deploying the keyring canvas in different sectors to provide more comprehensive insights. Such opportunities would provide empirical evidence from participants and partners to understand how the keyring metaphor may reveal opportunities for innovation.

The research team have also had previous success in using methods to segue blockchain thinking into designing with traditional fiat currencies (Murray-Rust et al., 2023), and at present the NFT method reported in this paper is orientated toward the assumption that blockchain technology is a prerequisite; future work may involve using the canvas to collaborate with organisations that are able to leverage open banking technology to design new products and services that use fiat currency rather than cryptocurrencies.

Although the research refers to the implicit challenges of working within platform economies in which money and data become entangled, future deployment with 'real-world' partners would allow us to better understand how the metaphor and canvas method would reveal ethical challenges. Such work could lead to an ethical framework for implementing non-fungible design, addressing privacy, consent, and potential unintended consequences.

Finally, the case studies featured do not touch upon the opportunities for policy. Within appropriate settings, with regional governments, the team are interested in understanding how the canvas could be used to understand how non-fungible design approaches might influence or require changes in policy and regulation, particularly in areas such as digital identity and data protection.

Acknowledgements

This research has been funded by a series of projects funded by the UK research councils: EPSRC Tales of Things and Electronic Memory project (EP/H007318/1), ESRC After Money project (ES/N007018/1), and the EPSRC DECaDE Centre (EP/T022485/1). We thank all participants and partners who took part in the research that informed our findings and insights.

References

Bandelj, N., Wherry, F. and Zelizer, V. (2017). *Money Talks: Explaining How Money Really Works*. Princeton: Princeton University Press. https://doi.org/10.1515/9781400885268.

Barad, K. (2003). "Posthumanist Performativity: Toward an Understanding of How Matter Comes to Matter", Signs: Journal of Women in Culture and Society 28 (3), pp. 801–31.

Bielefeld, S. (2018). Cashless Welfare Transfers for "Vulnerable" Welfare Recipients: Law, Ethics and Vulnerability. *Feminist Legal Studies*, *26*(1), 1–23. https://doi.org/10.1007/s10691-018-9363-6.

Bielefeld, S. (2021). Administrative burden and the Cashless Debit Card: Stripping time, autonomy, and dignity from social security recipients. *Australian Journal of Public Administration*, *80*(4), 891–911. https://doi.org/10.1111/1467-8500.12509.

de Jode, M.L., Barthel, R. and Hudson-Smith, A. (2011). Tales of things: the story so far. *Proceedings of the 2011 International Workshop on Networking and Object Memories for the Internet of Things*, 19–20. https://doi.org/10.1145/2029932.2029940.

Elsden, C., Manohar, A., Briggs, J., Harding, M., Speed, C. and Vines, J. (2018). Making Sense of Blockchain Applications: A Typology for HCI. *Proceedings of the 2018 CHI Conference on Human Factors in Computing Systems (CHI '18) New York, NY, USA,* Paper 458, 1–14. Association for Computing Machinery. https://doi.org/10.1145/3173574.3174032.

Elsden et al. (2024) A Token Gesture: Non-Transferable NFTs, Digital Possessions and Ownership Design. Forthcoming in PACM Journal (CSCW).

Elsden, C., Morgan, E., Tallyn, E., R. Black, S., Disley, M., Schafer, B., Murray-Rust, D. and Speed, C. (2024) A Token Gesture: Non-Transferable NFTs, Digital Possessions and Ownership Design. Proc. ACM Hum.-Comput. Interact. 8, CSCW1, Article 25 (April 2024), 29 pages. https://doi.org/10.1145/3637302.

Foth, M. (2017). The promise of blockchain technology for interaction design. *Proceedings of the 29th Australian Conference on Computer-Human Interaction*, 513–517. ACM. https://doi.org/10.1145/3152771.3156168.

Gaver, W., Beaver, J. and Benford, S. (2003). Ambiguity as a resource for design. *Proceedings of the SIGCHI Conference on Human Factors in Computing Systems (CHI '03) New York, NY, USA*, 233–40. Association for Computing Machinery. https://doi.org/10.1145/642611.642653.

Giaccardi, E., Speed, C. and Netten, M. (2016). Things2Things: Designing in the connected everyday. Delft, The Netherlands: Delft University of Technology.

Gruin, J. (2021). The epistemic evolution of market authority: Big data, blockchain and China's neostatist challenge to neoliberalism. *Competition & Change*, *25*(5), 580–604. https://doi.org/10.1177/1024529420965524.

Hart, K. (2001). *The memory bank: money in an unequal world.* London: Profile.

Huttunen, J. and Joutsenvirta, M. (2019). Monies, economies and democracy: cultivating ambivalence in the co-design of digital currencies. *CoDesign*, *15*(3), 228–42. 10.1080/15710882.2019.1631352.

Jia, L. (2020). Unpacking China's Social Credit System: Informatization, Regulatory Framework, and Market Dynamics. *Canadian Journal of Communication*, *45*(1). https://doi.org/10.22230/cjc.2020v45n1a3483.

MacKenzie, D. (2022). Spoofing: Law, materiality and boundary work in futures trading. *Economy and Society*, *51*(1), 1-22. 10.1080/03085147.2022.1987753.

Murray-Rust, D., Elsden, C., Nissen, B., Tallyn, E., Pschetz, L. and Speed, C. (2023) Blockchain and Beyond: Understanding Blockchains Through Prototypes and Public Engagement. ACM Trans. Comput.-Hum. Interact. 29, 5, Article 41 (October 2022), 73 pages. https://doi.org/10.1145/3503462.

Natanelov, V., Cao, S., Foth, M. and Dulleck, U. (2022). Blockchain Smart Contracts for Supply Chain Finance: Mapping the Innovation Potential in Australia-China Beef Supply Chains. *Journal of Industrial Information Integration*, 100389. https://doi.org/10.1016/j.jii.2022.100389.

Nissen, B., Pschetz, L., Murray-Rust, D., Mehrpouya, H., Oosthuizen, S. and Speed, C. (2018). GeoCoin: Supporting Ideation and Collaborative Design with Smart Contracts. *Proceedings of the 2018 CHI Conference on Human Factors in Computing Systems (CHI '18) New York, NY, USA*, Paper 163, 1–10. Association for Computing Machinery. https://doi.org/10.1145/3173574.3173737.

O'Dwyer, R. (2019). Cache society: transactional records, electronic money, and cultural resistance. *Journal of Cultural Economy*, *12*(2), 133–53, 10.1080/17530350.2018.1545243.

Speed, C. and Maxwell, R. (2015) Designing through value constellations. interactions 22, 5 (September-October 2015), 38–43. https://doi.org/10.1145/2807293.

Speed, C. (2012) Mobile Ouija Boards. In On Heritage. Edited by Giaccardi, E. London: Routledge. pp.179–96.

Swift (2023) ESG is set to shape the future of finance. Weblink: https://www.swift.com/news-events/news/esg-set-shape-future-finance Visited 14th January 2024.

Valeonti, F., Bikakis, A., Terras, M., Speed, C., Hudson-Smith, A. and Chalkias, K. (2021). Crypto Collectibles, Museum Funding and OpenGLAM: Challenges, Opportunities and the Potential of Non-Fungible Tokens (NFTs). *Applied Sciences*, *11*(21), 9931. https://doi.org/10.3390/app11219931.

Zelizer, V.A. (1989). The Social Meaning of Money: "Special Monies." *American Journal of Sociology*, *95*(2), 342–77. http://www.jstor.org/stable/2780903.

9 Evaluation of Strategic Resources in a Typical Short Food Supply Chain for Potential Blockchain Implementation

Stessy Coutelle[1*], Rami Alkhudary[2] and Pierre Féniès[3]

1. Introduction

Sustainable supply chain management focuses on the synchronization of material, information, and financial flows, while enhancing collaboration among stakeholders, particularly through shorter, localized supply chains (Kamble et al., 2020; Seuring and Müller, 2008). This approach supports sustainability by emphasizing the three foundational pillars of sustainable development: economic, environmental, and social objectives. These pillars are continuously adapted to respond to the evolving demands and expectations of consumers. While research into the benefits of shorter supply chains that prioritize local products is still emerging, it is important to deepen our understanding of their sustainability impacts (Bartoloni et al., 2022; Elkington, 1998; Paciarotti and Torregiani, 2021).

Short food supply chains are designed to fulfill the growing needs of consumers and producers who seek deeper personal connections, sustainable consumption, and a reduced ecological footprint, integrating information systems for better management (Malak-Rawlikowska et al., 2019; Thomé et al., 2021). Despite their benefits, these supply chains face increasing criticism, particularly concerning their transportation strategies. A major issue is the suboptimal use of transportation capacity, where instead of leveraging mass distribution benefits, food is often transported over unnecessarily long distances. Addressing this inefficiency is essential to enhance the long-term sustainability of local food networks, using advanced information systems to optimize routes and reduce redundancies (Martikainen et al., 2014).

Blockchain is often described as an information system that enhances operational efficiency (Alkhudary, Queiroz, et al., 2022; Kshetri, 2018). Its various functions work together to optimize operational processes by reducing transaction time and costs, promoting information sharing, ensuring data security, and building trust among supply chain actors (Lacity, 2022; Treiblmaier et al., 2021). Introducing a blockchain-based applications for farmers has the potential

[1-3] Université Paris-Panthéon-Assas, LARGEPA, Paris, France.
Email: rami.alkhudary@u-paris2.fr[2]; pierre.fenies@u-paris2.fr[3]
* Corresponding author: stessy.coutelle@u-paris2.fr

to streamline inventory management, transportation, and harvesting processes. This can lead to a reduction in food miles and food waste, moving towards more sustainable supply chains (Cao, Johnson, et al., 2023; Upadhyay et al., 2021). Moreover, this technological advancement fosters greater traceability and transparency and encourages the adoption of better business practices by making relevant information readily available to consumers (Cao, Johnson, et al., 2023; Cao, Xu, et al., 2023). However, most literature discussing blockchain in the food supply chain has not explicitly assessed the current state of strategic resources required to unlock the benefits of blockchain (Treiblmaier, 2018).

We aim to evaluate the strategic resources of a typical short food supply chain in Versailles, France, utilizing the resource-based view (RBV) framework (Barney, 1991; Duong and Paché, 2015). Our focus is to assess the current state of these resources to determine if they are sufficient to support the implementation of blockchain. The adoption of blockchain could improve traceability, offering a competitive advantage in markets increasingly focused on sustainability and the origins and quality of food (Cao, Xu, et al., 2023; Powell et al., 2022). We conduct an exploratory study to assess the strategic resources in a food supply chain in Versailles, typically an example of a short food supply chain. We then conceptually evaluate the readiness for blockchain implementation and consider the strategic changes necessary to utilize this technology. By taking this proactive approach, we aim to ensure that blockchain deployment is strategically aligned with enhancing supply chain collaboration and thereby generating a competitive edge in a dynamic market.

This chapter begins by providing an overview of the typical food supply chain. The discussion then shifts towards the potential for blockchain to foster a more sustainable food supply chain. The chapter further delves into the RBV framework. The methodology section outlines the research design and analytical methods employed. The evaluation section examines the current resources in a typical short food supply chain in Versailles and recommends some guidelines to meet the minimum requirements for blockchain implementation. Finally, the chapter concludes with summarizing remarks.

2. Food Supply Chain

The expansion of the global food industry is contributing to the increasing complexity of the sector, resulting in a higher number of stakeholders, longer transportation routes for perishable goods, and greater disparities in information dissemination within food supply chains. This growing complexity introduces a range of challenges at various stages of the supply chain, including issues such as food waste, safety concerns, and vulnerabilities related to costs (Alkhudary, Brusset, et al., 2022; Gokarn and Kuthambalayan, 2017). For example, over 30% of food is lost or wasted due to deficiencies within food supply networks (Porter and Reay, 2016). To address these challenges and enhance food supply chains, there is a growing focus on fostering increased collaboration (Barratt, 2004). This involves the establishment of diverse cooperative ventures and digital supply chains (Dubey et al., 2020; Dyer and Singh, 1998).

The emergence of digital technology is transforming consumers' long-term patterns of food sourcing and is also impacting the geographical distribution of market supply (Mashalah et al., 2022). The proliferation of various food distribution routes can be attributed to the wide range of consumer preferences. These include various retail channels, namely, supermarkets, small stores, home delivery and drive-through services, direct sales, and online retail. A range of supplementary distribution methods has arisen to accommodate the varying demands of consumers. To optimize the food supply chain and adapt to the industry's evolving dynamics, it is important to establish collaborative partnerships among the various stakeholders involved. This strategy not only improves the development of new products but also enhances the focus on customer needs and the value creation process (Rejeb et al., 2021; Salam, 2017).

There is a growing need to enhance collaboration among local food producers, especially in logistical operations. Effective collaboration is important for eliminating non-value-adding costs, optimizing resource utilization, and enhancing the supply chain. Despite the rising demand and interest in locally sourced food, local producers often find themselves sidelined within the traditional food supply chain. They face challenges including reduced production levels, limited resources, high prices, complexities in product development and marketing, difficulties in forming business partnerships, and risks associated with ensuring dependable delivery. These challenges underscore the critical need for innovation and adaptation in the food supply chain to better integrate and leverage the potential of local food producers through advanced information systems. These systems can play a key role in streamlining processes, improving communication, and facilitating more integration of local food sources into broader market ecosystems (Cao et al., 2021; Powell et al., 2022).

3. Towards a More Sustainable Food Supply Chain with Blockchain

Blockchain is a decentralized information system with the potential to revolutionize the operations of food supply chains (Behnke and Janssen, 2020; Kamilaris et al., 2019). This technology supports different models, namely, private, public, and consortium blockchains, each designed to meet specific needs (Al-Sumaidaee et al., 2023). It offers numerous advantages, such as reduced transaction costs and times, increased visibility, enhanced security measures, and the cultivation of trust within the supply chain (Alkhudary and Féniès, 2022; Cao, Xu, et al., 2023). Different industries are increasingly adopting blockchain applications within their logistical processes, recognizing the technology's potential to provide a competitive advantage (Al-Sumaidaee et al., 2021). This approach has been proven, particularly in the food industry (Brusset et al., 2024; Bumblauskas et al., 2020). The diverse applications of blockchain technology are used to meet the needs of customers and respond to the changing demands of the industry (Menon and Jain, 2022).

Sustainability in the food supply chain gained a lot of attention in recent years, addressing key environmental, social, and economic challenges. Research by Kazancoglu et al. (2021) and Yamoah and Yawson (2023) emphasizes the need for sustainable business practices to enhance resilience and minimize negative impacts, highlighting the role of external factors like climate change and the COVID-19 pandemic. Further studies by D. Li et al. (2014) and Malak-Rawlikowska et al. (2019) explore dimensions such as traceability, stakeholder collaboration, and sustainability metrics, addressing how to balance economic, environmental, and social factors. Additionally, works by Smith (2008) and Zhu et al. (2018) discuss the development and advancement of sustainable business practices, stressing the importance of continuous innovation for improving traceability and customer satisfaction. The body of literature underscores a comprehensive approach to sustainability in food supply chains to achieve long-term sustainability objectives (Cao, Johnson, et al., 2023; Menozzi, 2014).

Li et al. (2023) underscore the importance of blockchain in addressing the three fundamental aspects of sustainability: economic, social, and environmental. For example, it plays a key role in advancing economic sustainability by reducing costs associated with food waste and product recalls (Schmidt and Wagner, 2019). This reduction is made possible through the traceability, transparency, and immutability features of blockchain technology (Longo et al., 2019). Additionally, smart packaging enabled by blockchain technology provides precise data on the state of food items, further reducing avoidable disposals. Achieving social sustainability within food supply chains requires addressing consumer concerns regarding food products. Blockchain mitigates these concerns through its transparency and traceability features, providing insights into food origin, quality, and environmental impact. Blockchain also aligns with fair trade principles, offering consumers transparency about how much of their payment benefits the farmer. An example is the use of blockchain technology to combat forced labor in the sugarcane industry and to trace the ethical sourcing of products like coffee, enhancing customer awareness (Pollock, 2020; Thiruchelvam et al., 2018).

Blockchain contributes to environmental sustainability by reducing greenhouse gas emissions through precise monitoring of environmental data (Guo et al., 2020; Rane and Thakker, 2019). Blockchain enhances compliance with environmental regulations, involving regulated entities and the public, thus preventing corruption and maladministration. Moreover, the study conducted by Casino et al. (2021) highlights the increasing consumer demand for transparency, especially in terms of food safety. Another example, Behnke and Janssen (2020) explore the application of blockchain in tracing items and creating transparency in procurement processes. However, it is important to note that certain constraints are associated with blockchain technology, particularly in terms of food traceability. These constraints often relate to regulatory requirements, supply chain structuring, and manufacturing process optimization. Standardized master data and interfaces are also lacking, limiting complete automation and underscoring the need for better supply chain structuring before blockchain implementation.

4. Resource-Based View Theory

The emergence of resource theory, also known as the Resource-Based View (RBV), can be seen as a response to the positioning school of thought, which placed an emphasis on an organization's strategic positioning (Porter, 1985). The RBV theory argues that a firm's competitive advantage derives from a specific subset of its resources, and even a smaller fraction of these resources can result in long-term superior performance (Barney, 1991; Teece et al., 1997). Competitive advantage is achieved by holding strategic resources, which can be sustained as long as the company can protect them and no viable alternatives emerge (Halldorsson et al., 2007; Wernerfelt, 1984). The use of RBV raises several concerns, including the competitive implications of reallocating resources within the blockchain domain and the need for additional resources to deploy this technology (Treiblmaier, 2018).

According to Martinez et al. (2019), blockchain is perceived as a dynamic capability within the context of supply chain management. The authors highlight that this technology has tangible effects through its application in industry-specific settings. The implementation of blockchain technology leads to improved process efficiency by reducing the number of activities, shortening the average order processing time, decreasing effort, ensuring traceability of orders, and enhancing visibility for different parties involved in the supply chain. Gökalp et al. (2022) conducted a systematic literature review to examine the factors influencing the adoption of blockchain technology in the context of supply chain management. Their findings indicate that factors related to the environment have a greater impact than those related to technology or organization. The authors emphasize the significance of improvements in information and communication technology (ICT) as a potential driver of competitive advantage.

The RBV framework provides a comprehensive understanding of how businesses can enhance their competitive advantage by effectively adopting the necessary capabilities, skills, procedures, technology, and networks required for the successful implementation of digital platforms (J. Aslam et al., 2023). When examining blockchain from the RBV perspective, it becomes evident that it is important to evaluate not only the fundamental resources and capabilities necessary for the utilization of this technology but also to comprehend the interrelated skills and capabilities associated with it. The incorporation of blockchain technology within the realm of client order administration contributes to the improvement of traceability and visibility, thereby establishing a competitive edge (Chillakuri and Attili, 2022).

Strategic resources in the RBV framework applied to supply chain management are identified as tangible, intangible, knowledge and strategic capabilities (Brusset and Teller, 2017). These resources, cited as valuable, rare, irreplaceable, and challenging to replicate, are foundational to gaining and maintaining a competitive edge (Barney, 1991). Tangible assets like physical infrastructure, technological tools, and financial assets are important for value creation in supply chains (Qrunfleh and Tarafdar, 2013). On the other hand, intangible assets and knowledge elements also play a key role in the RBV framework for supply chains (Bratianu, 2015; Hult et al., 2006). Effective integration of knowledge assets, expertise, and information management practices can enhance decision-making and increase operational efficiency, thereby yielding competitive advantage (Cepeda and Vera, 2007; Grimaldi et al., 2017; Nielsen, 2006; Steininger et al., 2021). Capabilities, which include the necessary skills, processes, and routines are critical strategic resources within the RBV framework (Davies and Brady, 2000). These capabilities are instrumental in enhancing a firm's strategic position and performance in the supply chain (H. Aslam et al., 2018).

Research suggests that the RBV defines organizations as diverse collections of resources, i.e., assets and capabilities, that can be used for strategic value creation (Wang et al., 2022). Applying this framework to the food supply chain in Versailles, we explore the strategic resources available to determine if they are sufficient to support the implementation of blockchain. This assessment is crucial as blockchain can potentially enhance traceability, transparency, and trust among stakeholders in the supply chain (Brusset et al., 2024). The RBV framework assists in systematically identifying these resources to evaluate their readiness and adequacy for adopting new technologies like blockchain (Martinez et al., 2019).

5. Methodology

The study's research methodology follows a case study approach (Barratt et al., 2011; Eisenhardt, 1989; McCutcheon and Meredith, 1993). Precisely, we examine the complicated dynamics of the food supply chain in the Plaine de Versailles region of Île-de-France, France. Our study employed a rigorous selection of participants that included a wide range of individuals who play critical roles within the network of the food supply chain in the Plaine de Versailles. The inclusion of various stakeholders in this study ensured that the business practices were represented to provide a comprehensive understanding of the operational landscape. To enhance the reliability and precision of our findings, we adhered to the principle of data triangulation (Yin, 2017). These methods included conducting in-depth semi-structured interviews with key stakeholders (Table 9.1). The interviews are guided by an interview protocol (see appendix). Notably, one of the authors of this paper immersed themselves in the Plaine de Versailles association for a two-month period in 2023 to conduct a close investigation of this food supply chain.

To deepen our findings and provide a more comprehensive framework for the case study, we conducted a secondary data analysis (Tate and Happ, 2018; Ziebland and Hunt, 2014). A wide range of secondary sources, including annual reports, professional journals, websites, and social networks, were examined to extend and validate the findings from the primary case study. An example of the secondary sources is the website of the Plaine de Versailles association, which

Table 9.1 Selection of interview participants

Interviewees	*Function*	*Number of participants*	*Duration of interview*
Producer	President of a cooperative of 30 producers.	2	90 min
Producer	Producer and member of the association Plaine de Versailles.	1	60 min
Wholesaler	Director, wholesaler for various customer sectors: 800 products, and 12 local producers.	1	60 min
Wholesaler	Director of Operations: 4500 products, and 285 partners.	1	60 min
Matchmaking platform	Co-founder of a platform whose mission is to make local and direct commerce as simple as centralized commerce.	1	60 min

shares monthly blogs and reports on the food supply chain in the region. Integrating primary and secondary data sources into our study improved its scope and resulted in a synergistic approach. In this way, the study aims to provide valuable insights into the challenges and complexities faced by supply chain actors in an era of technological advances and changing consumer preferences.

In the initial phase of the data analysis, transcripts from the interviews and other secondary data were collected and subjected to a rigorous verbatim coding process. This technique was aligned with an analytical framework based on the strategic resource view, to uncover and categorize strategic assets within an organization or ecosystem. Analyzing qualitative content based on previously established research questions or frameworks is a common method that follows the guidelines of Wolfswinkel et al. (2013). The analysis facilitated a deeper understanding of the dynamics and challenges faced by various stakeholders. Moreover, the coding process involved identifying, analyzing, and reporting patterns within the data (Yin, 2017), which provided insights into the interactions between stakeholders and their impact on the food supply chain.

We utilized the RBV framework to evaluate the strategic resources within a typical short food supply chain in Versailles and to determine if the existing resources are sufficient for implementing blockchain technology to enhance traceability, sustainability, and competitive advantage. These resources are categorized into five dimensions: technological, organizational, physical, relational, and expertise resources (Duong and Paché, 2015). Technological resources (Fung, 2019) encompass the necessary communication systems and internet technologies. Organizational resources (Windsperger et al., 2018) include the shared values and inter-organizational coordination. Physical resources (Turulja and Bajgoric, 2016) cover the means of storage, transport, and distribution, as well as the production and packaging capabilities required to handle the produce. Relational resources (Dyer and Singh, 1998) refer to the common projects, the duration of relationships, and the sharing of reliable information, all of which foster strong partnerships and collaboration. Lastly, expertise resources (Helfat and Peteraf, 2003) pertain to the quality of production and the knowledge, training, and expertise that the organization possesses, ensuring high standards and continuous improvement.

6. Evaluation of Strategic Resources for Blockchain Implementation

The food supply chain in the Plaine de Versailles region comprises four main stages: delivery, production, distribution, and sales. However, the distinct characteristics of this supply chain are evident in the diverse range of tactics and activities implemented to manage its operations. These activities often lack regularity and cohesiveness due to the multitude of factors involved. Growers operate independently, pursuing their specific agendas with minimal communication between different operational areas. This isolation poses challenges in terms of optimizing resources and enhancing operations. Some producers collaborate with external processing service providers, involving themselves in several processing stages. While this approach has improved the supply chain, it also introduces challenges related to contractual obligations and the need for coordination among various stakeholders.

Expertise seems to be the primary source of competitive advantage. Producers focus on agricultural yields and diverse cultivation methods. Some explore lesser-used techniques and concentrate on mastering their land, occasionally in defiance of guidelines. For example, one producer shared that they cultivated crops to lessen their ecological footprint, opting for products not typically emphasized to consumers. This requires the identification of new distribution channels to market produce critical for the sustainability of agricultural lands.

Regarding relational resources, producers encounter challenges when trying to integrate into conventional distribution channels, often due to insufficient logistical development. Although they hold shared values, their organizational resources are restricted, and their physical resources mostly involve the agricultural land they operate. Some producers utilize physical resources by organizing events, establishing pick-your-own options, or maintaining on-site sales locations. Many point to time constraints as this activity often diverges from their primary business, particularly in the case of large-scale agriculture.

The wholesaler's contribution to the supply chain stems from their capability to utilize organizational resources despite forecasting challenges for human resources. They can streamline and coordinate their logistics operations and manage various volumes across a regional scale. A noted limitation is physical resources. Precisely, they have storage space that is available yet constrained due to the urban-agricultural land setting. They maintain operational logistics skills but lack sensitivity to consumer expectations, especially concerning the traceability of products. Distributors attempt to leverage various technological resources, particularly through new platforms that aggregate orders for local products. However, their organizational structure often hinders the activation of these resources in direct sales channels. Small producers have contributed to mapping out the distribution of strategic resources among various stakeholders, aligned with the traceability of physical flows.

The deployment of a blockchain application on the scale of a territorial association can generate new strategic skills for producers. Blockchain can reduce intermediaries in short supply chains, offer transaction records, and streamline information exchanges, particularly with distributors with whom they cannot communicate smoothly. This would also allow new producers to focus on their yield and core business while acquiring skills from certain networking platforms. Blockchain can facilitate exchanges with various actors, such as wholesalers, large retail chains, and end customers, thus achieving a unique information management system. Stakeholders can manage their volume and production, establish detailed forecasts which, in the long term, would enable value exchange. Blockchain can also improve trust and reduce the multiplication of flows and the integration of various logistical processes.

After evaluating the strategic resources of a typical short food supply chain in Versailles, we propose three steps to develop a blockchain project. First, it is important to clearly define the requirements of a new information system for each stakeholder. Additionally, establishing the technology's use cases, the types of data exchanged, and the consensus among stakeholders is necessary. We recommend documenting the processes and contracts tailored to the environment, the roles of different actors within the supply chain, and the need for all stakeholders to commit at the network's first layer when deploying blockchain. Second, determining the number of nodes, network layers, and the nature and number of smart contracts to deploy is essential. These steps should be defined in collaboration with supply chain stakeholders and blockchain project managers before moving on to the design and deployment of the application.

Finally, recruiting a team of blockchain developers is one of the more challenging tasks, given the ongoing development of the technology and the difficulty of finding skilled personnel in the computer development market. Developers then write code that meets the organization's needs. Unlike typical information system development projects, conducting a proof-of-concept (POC) is important for testing and refining the system. This phase includes managing system connectivity and integration, maintenance, and updates to ensure network security and mitigate IT threats. It is also important to mobilize expertise not only to deploy the technology but also to engage stakeholders by making application layers user-friendly and familiarizing them with the technology to optimize performance utilization.

Conclusion

The food supply chain in Versailles serves as an example of a typical short food supply chain, showcasing the complex dynamics that define modern supply chains. However, it is important to acknowledge that the food supply chain, like various other industries, faces resource constraints that hinder its capacity to utilize technologies such as blockchain. The findings underscore the urgent need for increased collaboration among supply chain stakeholders to overcome the fragmentation of information systems. Growers should thoughtfully tailor their logistical strategies to align with their specific circumstances and objectives. To meet the expectations of an informed consumer base, it is important for producers, wholesalers, and other stakeholders to adapt their approaches, prioritizing transparency, traceability, and sustainability in their business practices. A blockchain system can help the food supply chain become more dynamic and respond to market needs. However, a minimum requirement of strategic resources is necessary to utilize this technology and reap its benefits. These observations lay the groundwork for future research efforts and offer valuable insights into the constraints and possibilities within food supply chains.

This study is not without limitations. First, the study focuses on a specific geographical area, limiting the generalizability of the findings to other regions with different supply chain dynamics and resource availability. Second, the reliance on qualitative data, such as interviews with a small number of stakeholders, may introduce subjectivity and bias, potentially affecting the reliability of the analysis. Additionally, the rapid evolution of blockchain technology and its applications means that the study's findings may quickly become outdated. The chapter also does not account for potential regulatory changes that could impact the implementation of blockchain in food supply chains. Lastly, while the study highlights the need for strategic resources, it does not provide detailed implementation strategies.

References

Alkhudary, R., Brusset, X., Naseraldin, H. and Féniès, P. (2022). Enhancing the competitive advantage via Blockchain : An olive oil case study. *IFAC-PapersOnLine*, *55*(2), 469–74. https://doi.org/10.1016/j.ifacol.2022.04.238.

Alkhudary, R. and Féniès, P. (2022). Blockchain and Trust in Supply Chain Management : A Conceptual Framework. *IFAC-PapersOnLine*, *55*(10), 2402–2406. https://doi.org/10.1016/j.ifacol.2022.10.068.

Alkhudary, R., Queiroz, M.M. and Féniès, P. (2022). Mitigating the risk of specific supply chain disruptions through blockchain technology. *Supply Chain Forum: An International Journal*, 1-11. https://doi.org/10.1080/16258312.2022.2090273.

Al-Sumaidaee, G., Alkhudary, R., Zilic, Z. and Féniès, P. (2021). A blueprint towards an integrated healthcare information system through blockchain technology. HEALTHINFO 2021, The Sixth International Conference on Informatics and Assistive Technologies for Health-Care, Medical Support and Wellbeing.

Al-Sumaidaee, G., Alkhudary, R., Zilic, Z. and Féniès, P. (2023). Configuring Blockchain Architectures and Consensus Mechanisms : The Healthcare Supply Chain as a Use Case. In A. Bouras, I. Khalil, & B. Aouni (Éds.), *Blockchain Driven Supply Chains and Enterprise Information Systems* (pp. 135-150). Springer International Publishing. https://doi.org/10.1007/978-3-030-96154-1_7.

Aslam, H., Blome, C., Roscoe, S. and Azhar, T.M. (2018). Dynamic supply chain capabilities : How market sensing, supply chain agility and adaptability affect supply chain ambidexterity. *International Journal of Operations & Production Management*, *38*(12), 2266-85. https://doi.org/10.1108/IJOPM-09-2017-0555.

Aslam, J., Saleem, A. and Kim, Y. B. (2023). Blockchain-enabled supply chain management : Integrated impact on firm performance and robustness capabilities. *Business Process Management Journal*, *29*(6), 1680–1705. https://doi.org/10.1108/BPMJ-03-2023-0165.

Barney, J. (1991). Firm Resources and Sustained Competitive Advantage. *Journal of Management*, *17*(1), 99–120. https://doi.org/10.1177/014920639101700108.

Barratt, M. (2004). Understanding the meaning of collaboration in the supply chain. *Supply Chain Management: An International Journal*, *9*(1), 30–42. https://doi.org/10.1108/13598540410517566.

Barratt, M., Choi, T.Y. and Li, M. (2011). Qualitative case studies in operations management : Trends, research outcomes, and future research implications. *Journal of Operations Management*, *29*(4), 329–342. https://doi.org/10.1016/j.jom.2010.06.002.

Bartoloni, S., Ietto, B. and Pascucci, F. (2022). Do connoisseur consumers care about sustainability? Exploring coffee consumption practices through netnography. *British Food Journal*, *124*(13), 305–321. https://doi.org/10.1108/BFJ-07-2021-0814.

Behnke, K. and Janssen, M.F.W.H.A. (2020). Boundary conditions for traceability in food supply chains using blockchain technology. *International Journal of Information Management*, *52*, 101969. https://doi.org/10.1016/j.ijinfomgt.2019.05.025.

Bratianu, C. (2015). *Organizational Knowledge Dynamics : Managing Knowledge Creation, Acquisition, Sharing, and Transformation*. IGI Global. https://doi.org/10.4018/978-1-4666-8318-1.

Brusset, X., Kinra, A., Naseraldin, H. and Alkhudary, R. (2024). Increasing willingness to pay in the food supply chain : A blockchain-oriented trust approach. *International Journal of Production Research*, 1–22. https://doi.org/10.1080/00207543.2024.2352763.

Brusset, X. and Teller, C. (2017). Supply chain capabilities, risks, and resilience. *International Journal of Production Economics*, *184*, 59–68. https://doi.org/10.1016/j.ijpe.2016.09.008.

Bumblauskas, D., Mann, A., Dugan, B. and Rittmer, J. (2020). A blockchain use case in food distribution : Do you know where your food has been? *International Journal of Information Management*, *52*, 102008. https://doi.org/10.1016/j.ijinfomgt.2019.09.004.

Cao, S., Johnson, H. and Tulloch, A. (2023). Exploring blockchain-based Traceability for Food Supply Chain Sustainability : Towards a Better Way of Sustainability Communication with Consumers. *Procedia Computer Science*, *217*, 1437–45. https://doi.org/10.1016/j.procs.2022.12.342.

Cao, S., Powell, W., Foth, M., Natanelov, V., Miller, T. and Dulleck, U. (2021). Strengthening consumer trust in beef supply chain traceability with a blockchain-based human-machine reconcile mechanism. *Computers and Electronics in Agriculture*, *180*, 105886. https://doi.org/10.1016/j.compag.2020.105886.

Cao, S., Xu, H. and Bryceson, K.P. (2023). Blockchain Traceability for Sustainability Communication in Food Supply Chains : An Architectural Framework, Design Pathway and Considerations. *Sustainability*, *15*(18), 13486. https://doi.org/10.3390/su151813486.

Casino, F., Kanakaris, V., Dasaklis, T.K., Moschuris, S., Stachtiaris, S., Pagoni, M. and Rachaniotis, N.P. (2021). Blockchain-based food supply chain traceability : A case study in the dairy sector. *International Journal of Production Research*, *59*(19), 5758–70. https://doi.org/10.1080/00207543.2020.1789238.

Cepeda, G. and Vera, D. (2007). Dynamic capabilities and operational capabilities : A knowledge management perspective. *Journal of Business Research*, *60*(5), 426–37. https://doi.org/10.1016/j.jbusres.2007.01.013.

Chillakuri, B. and Attili, V.S.P. (2022). Role of blockchain in HR's response to new-normal. *International Journal of Organizational Analysis*, *30*(6), 1359–78. https://doi.org/10.1108/IJOA-08-2020-2363.

Davies, A. and Brady, T. (2000). Organisational capabilities and learning in complex product systems : Towards repeatable solutions. *Research Policy*, *29*(7-8), 931–53. https://doi.org/10.1016/S0048-7333(00)00113-X.

Dubey, R., Gunasekaran, A., Bryde, D.J., Dwivedi, Y.K. and Papadopoulos, T. (2020). Blockchain technology for enhancing swift-trust, collaboration and resilience within a humanitarian supply chain setting. *International Journal of Production Research*, *58*(11), 3381–98. https://doi.org/10.1080/00207543.2020.1722860.

Duong, H.T. and Paché, G. (2015). Théorie des ressources appliquée à la logistique : Une identification de cinq dimensions clés. *Logistique & Management*, *23*(2), 55–72. https://doi.org/10.1080/12507970.2015.11673824.

Dyer, J.H. and Singh, H. (1998). The Relational View : Cooperative Strategy and Sources of Interorganizational Competitive Advantage. *The Academy of Management Review*, *23*(4), 660. https://doi.org/10.2307/259056.

Eisenhardt, K.M. (1989). Building Theories from Case Study Research. *The Academy of Management Review*, *14*(4), 532. https://doi.org/10.2307/258557.

Elkington, J. (1998). Partnerships from cannibals with forks : The triple bottom line of 21st-century business. *Environmental Quality Management*, *8*(1), 37–51. https://doi.org/10.1002/tqem.3310080106.

Fung, M.K. (2019). Fraudulent Financial Reporting and Technological Capability in the Information Technology Sector : A Resource-Based Perspective. *Journal of Business Ethics*, *156*(2), 577–89. https://doi.org/10.1007/s10551-017-3605-4.

Glaser, B.G. and Strauss, A.L. (1967). The discovery of grounded theory : Strategies for qualitative research.

Gökalp, E., Gökalp, M.O. and Çoban, S. (2022). Blockchain-Based Supply Chain Management : Understanding the Determinants of Adoption in the Context of Organizations. *Information Systems Management*, *39*(2), 100–121. https://doi.org/10.1080/10580530.2020.1812014.

Gokarn, S. and Kuthambalayan, T.S. (2017). Analysis of challenges inhibiting the reduction of waste in food supply chain. *Journal of Cleaner Production*, *168*, 595-604. https://doi.org/10.1016/j.jclepro.2017.09.028.

Grimaldi, M., Corvello, V., De Mauro, A. and Scarmozzino, E. (2017). A systematic literature review on intangible assets and open innovation. *Knowledge Management Research & Practice*, *15*(1), 90–100. https://doi.org/10.1057/s41275-016-0041-7.

Guo, S., Sun, X. and Lam, H.K.S. (2020). Applications of Blockchain Technology in Sustainable Fashion Supply Chains : Operational Transparency and Environmental Efforts. *IEEE Transactions on Engineering Management*, 1–17. https://doi.org/10.1109/TEM.2020.3034216.

Halldorsson, A., Kotzab, H., Mikkola, J.H. and Skjøtt-Larsen, T. (2007). Complementary theories to supply chain management. *Supply Chain Management: An International Journal*, *12*(4), 284–96. https://doi.org/10.1108/13598540710759808.

Helfat, C.E. and Peteraf, M.A. (2003). The dynamic resource-based view : Capability lifecycles. *Strategic Management Journal*, *24*(10), 997–1010. https://doi.org/10.1002/smj.332.

Hult, G.T.M., Ketchen, D.J., Cavusgil, S.T. and Calantone, R.J. (2006). Knowledge as a strategic resource in supply chains. *Journal of Operations Management*, *24*(5), 458–75. https://doi.org/10.1016/j.jom.2005.11.009.

Kamble, S.S., Gunasekaran, A. and Sharma, R. (2020). Modeling the blockchain enabled traceability in agriculture supply chain. *International Journal of Information Management*, *52*, 101967. https://doi.org/10.1016/j.ijinfomgt.2019.05.023.

Kamilaris, A., Fonts, A. and Prenafeta-Boldύ, F.X. (2019). The rise of blockchain technology in agriculture and food supply chains. *Trends in Food Science & Technology*, *91*, 640–52. https://doi.org/10.1016/j.tifs.2019.07.034.

Kazancoglu, Y., Ozbiltekin-Pala, M., Sezer, M.D., Ekren, B.Y. and Kumar, V. (2021). Assessing the Impact of COVID-19 on Sustainable Food Supply Chains. *Sustainability*, *14*(1), 143. https://doi.org/10.3390/su14010143.

Kshetri, N. (2018). 1 Blockchain's roles in meeting key supply chain management objectives. *International Journal of Information Management*, *39*, 80–89. https://doi.org/10.1016/j.ijinfomgt.2017.12.005.

Lacity, M.C. (2022). Blockchain : From Bitcoin to the Internet of Value and beyond. *Journal of Information Technology*, 026839622210863. https://doi.org/10.1177/02683962221086300.

Li, D., Wang, X., Chan, H.K. and Manzini, R. (2014). Sustainable food supply chain management. *International Journal of Production Economics*, *152*, 1-8. https://doi.org/10.1016/j.ijpe.2014.04.003.

Li, K., Lee, J.-Y. and Gharehgozli, A. (2023). Blockchain in food supply chains : A literature review and synthesis analysis of platforms, benefits and challenges. *International Journal of Production Research*, *61*(11), 3527–46. https://doi.org/10.1080/00207543.2021.1970849.

Longo, F., Nicoletti, L. and Padovano, A. (2019). Estimating the Impact of Blockchain Adoption in the Food Processing Industry and Supply Chain. *International Journal of Food Engineering*, 16(5-6). https://doi.org/10.1515/ijfe-2019-0109.

Malak-Rawlikowska, A., Majewski, E., Wąs, A., Borgen, S.O., Csillag, P., Donati, M., Freeman, R., Hoàng, V., Lecoeur, J.-L., Mancini, M.C., Nguyen, A., Saïdi, M., Tocco, B., Török, Á., Veneziani, M., Vittersø, G. and Wavresky, P. (2019). Measuring the Economic, Environmental, and Social Sustainability of Short Food Supply Chains. *Sustainability*, *11*(15), 4004. https://doi.org/10.3390/su11154004.

Martikainen, A., Niemi, P. and Pekkanen, P. (2014). Developing a service offering for a logistical service provider—Case of local food supply chain. *International Journal of Production Economics*, *157*, 318–26. https://doi.org/10.1016/j.ijpe.2013.05.026.

Martinez, V., Zhao, M., Blujdea, C., Han, X., Neely, A. and Albores, P. (2019). Blockchain-driven customer order management. *International Journal of Operations & Production Management*, *39*(6/7/8), 993-1022. https://doi.org/10.1108/IJOPM-01-2019-0100.

Mashalah, H.A., Hassini, E., Gunasekaran, A. and Bhatt (Mishra), D. (2022). The impact of digital transformation on supply chains through e-commerce : Literature review and a conceptual framework. *Transportation Research Part E: Logistics and Transportation Review*, *165*, 102837. https://doi.org/10.1016/j.tre.2022.102837.

McCutcheon, D.M. and Meredith, J.R. (1993). Conducting case study research in operations management. *Journal of Operations Management*, *11*(3), 239–256. https://doi.org/10.1016/0272-6963(93)90002-7.

Menon, S. and Jain, K. (2022). Blockchain Technology for Transparency in Agri-Food Supply Chain : Use Cases, Limitations, and Future Directions. *IEEE Transactions on Engineering Management*, 1–15. https://doi.org/10.1109/TEM.2021.3110903.

Menozzi, D. (2014). Extra-virgin olive oil production sustainability in northern Italy : A preliminary study. *British Food Journal*, *116*(12), 1942–59. https://doi.org/10.1108/BFJ-06-2013-0141.

Nielsen, P.A. (2006). Understanding dynamic capabilities through knowledge management. *Journal of Knowledge Management*, *10*(4), 59–71. https://doi.org/10.1108/13673270610679363.

Paciarotti, C. and Torregiani, F. (2021). The logistics of the short food supply chain : A literature review. *Sustainable Production and Consumption*, *26*, 428–42. https://doi.org/10.1016/j.spc.2020.10.002.

Pollock, D. (2020). Nestlé Expands Use Of IBM Food Trust Blockchain To Its Zoégas Coffee Brand. *Forbes*. https://www.forbes.com/sites/darrynpollock/2020/04/15/nestl-expands-use-of-ibm-food-trust-blockchain-to-its-zogas-coffee-brand/.

Porter, M.E. (1985). *Competitive Advantage : Creating and Sustaining Superior Performance* (First Printing). The Free Press.

Porter, S.D. and Reay, D.S. (2016). Addressing food supply chain and consumption inefficiencies : Potential for climate change mitigation. *Regional Environmental Change*, *16*(8), 2279–90. https://doi.org/10.1007/s10113-015-0783-4.

Powell, W., Foth, M., Cao, S. and Natanelov, V. (2022). Garbage in garbage out : The precarious link between IoT and blockchain in food supply chains. *Journal of Industrial Information Integration, 25*, 100261. https://doi.org/10.1016/j.jii.2021.100261.

Qrunfleh, S. and Tarafdar, M. (2013). Lean and agile supply chain strategies and supply chain responsiveness : The role of strategic supplier partnership and postponement. *Supply Chain Management: An International Journal, 18*(6), 571–82. https://doi.org/10.1108/SCM-01-2013-0015.

Rane, S.B. and Thakker, S.V. (2019). Green procurement process model based on blockchain–IoT integrated architecture for a sustainable business. *Management of Environmental Quality: An International Journal, 31*(3), 741–63. https://doi.org/10.1108/MEQ-06-2019-0136.

Rejeb, A., Keogh, J.G., Simske, S.J., Stafford, T. and Treiblmaier, H. (2021). Potentials of blockchain technologies for supply chain collaboration : A conceptual framework. *The International Journal of Logistics Management, 32*(3), 973–94. https://doi.org/10.1108/IJLM-02-2020-0098.

Salam, M.A. (2017). The mediating role of supply chain collaboration on the relationship between technology, trust and operational performance : An empirical investigation. *Benchmarking: An International Journal, 24*(2), 298–317. https://doi.org/10.1108/BIJ-07-2015-0075.

Schmidt, C.G. and Wagner, S.M. (2019). Blockchain and supply chain relations : A transaction cost theory perspective. *Journal of Purchasing and Supply Management, 25*(4), 100552. https://doi.org/10.1016/j.pursup.2019.100552.

Seuring, S. and Müller, M. (2008). From a literature review to a conceptual framework for sustainable supply chain management. *Journal of Cleaner Production, 16*(15), 1699–1710. https://doi.org/10.1016/j.jclepro.2008.04.020.

Smith, B.G. (2008). Developing sustainable food supply chains. Philosophical Transactions of the Royal Society B: *Biological Sciences, 363*(1492), 849–61. https://doi.org/10.1098/rstb.2007.2187.

Steininger, D., Mikalef, P., Pateli, A. and Guinea, A. (2021). Dynamic Capabilities in Information Systems Research : A Critical Review, Synthesis of Current Knowledge, and Recommendations for Future Research. *Journal of the Association for Information Systems*, forthcoming.

Tate, J.A. and Happ, M.B. (2018). Qualitative Secondary Analysis : A Case Exemplar. *Journal of Pediatric Health Care, 32*(3), 308–12. https://doi.org/10.1016/j.pedhc.2017.09.007.

Teece, D.J., Pisano, G. and Shuen, A. (1997). Dynamic Capabilities and Strategic Management. *Strategic Management Journal, 18*(7), 509–33.

Thiruchelvam, V., Mughisha, A.S., Shahpasand, M. and Bamiah, M. (2018). Blockchain-based Technology in the Coffee Supply Chain Trade : Case of Burundi Coffee. *Journal of Telecommunication, Electronic and Computer Engineering, 10*(3), 121–125.

Thomé, K.M., Cappellesso, G., Ramos, E.L.A. and Duarte, S.C. de L. (2021). Food Supply Chains and Short Food Supply Chains : Coexistence conceptual framework. *Journal of Cleaner Production, 278*, 123207. https://doi.org/10.1016/j.jclepro.2020.123207.

Treiblmaier, H. (2018). The impact of the blockchain on the supply chain : A theory-based research framework and a call for action. *Supply Chain Management: An International Journal, 23*(6), 545–59. https://doi.org/10.1108/SCM-01-2018-0029.

Treiblmaier, H., Rejeb, A., van Hoek, R. and Lacity, M. (2021). Intra- and Interorganizational Barriers to Blockchain Adoption : A General Assessment and Coping Strategies in the Agrifood Industry. *Logistics, 5*(4), 87. https://doi.org/10.3390/logistics5040087.

Turulja, L. and Bajgoric, N. (2016). Innovation and Information Technology Capability as Antecedents of Firms' Success. *Interdisciplinary Description of Complex Systems, 14*(2), 148–56. https://doi.org/10.7906/indecs.14.2.4.

Upadhyay, A., Mukhuty, S., Kumar, V. and Kazancoglu, Y. (2021). Blockchain technology and the circular economy : Implications for sustainability and social responsibility. *Journal of Cleaner Production, 293*, 126130. https://doi.org/10.1016/j.jclepro.2021.126130.

Wang, W., Lumineau, F. and Schilke, O. (2022). *Blockchains : Strategic Implications for Contracting, Trust, and Organizational Design* (1re éd.). Cambridge University Press. https://doi.org/10.1017/9781009057707.

Wernerfelt, B. (1984). A resource-based view of the firm. *Strategic Management Journal*, *5*(2), 171–80. https://doi.org/10.1002/smj.4250050207.

Windsperger, J., Hendrikse, G.W.J., Cliquet, G. and Ehrmann, T. (2018). Governance and strategy of entrepreneurial networks : An introduction. *Small Business Economics*, *50*(4), 671–76. https://doi.org/10.1007/s11187-017-9888-0.

Wolfswinkel, J.F., Furtmueller, E. and Wilderom, C.P.M. (2013). Using grounded theory as a method for rigorously reviewing literature. *European Journal of Information Systems*, *22*(1), 45–55. https://doi.org/10.1057/ejis.2011.51.

Yamoah, F.A. and Yawson, D. E. (2023). Special Issue : Sustainable Food Supply Chain Research. *Sustainability*, *15*(6), 4737. https://doi.org/10.3390/su15064737.

Yin, R.K. (2017). *Case Study Research and Applications*. SAGE Publications Inc. https://us.sagepub.com/en-us/nam/case-study-research-and-applications/book250150.

Zhu, Z., Chu, F., Dolgui, A., Chu, C., Zhou, W. and Piramuthu, S. (2018). Recent advances and opportunities in sustainable food supply chain : A model-oriented review. *International Journal of Production Research*, *56*(17), 5700–22. https://doi.org/10.1080/00207543.2018.1425014.

Ziebland, S. and Hunt, K. (2014). Using secondary analysis of qualitative data of patient experiences of health care to inform health services research and policy. *Journal of Health Services Research & Policy*, *19*(3), 177–82. https://doi.org/10.1177/1355819614524187.

Appendix | Interview Protocol

Introduction

One of the authors briefly explained the study's purpose, ensured confidentiality, and obtained consent to record and share data in an academic publication.

Background

Describe your role and responsibilities.
How long have you been involved in the Plaine de Versailles food supply chain?
What are the key logistical challenges you face?
Describe the current methods for tracking and managing inventory and transportation.
What are the main communication channels among stakeholders?

Resource Evaluation (RBV Framework)

What technologies do you currently use?
How effective are these technologies?
What physical resources do you have?
Are there any constraints with these resources?
Describe your relationships with other stakeholders.
How do these relationships impact the supply chain?
What expertise and knowledge do you and your team possess?
What is your understanding of blockchain technology?
What benefits do you see in adopting blockchain, such as enhanced sustainability?
What obstacles do you foresee in implementing blockchain?

10 | Circular Economy through Blockchain and Data Analytics

Behzad Esmaeilian[1*], Hao-yu Liao[2] and Sara Behdad[3]

1. Introduction

In recent years, a major change has taken place in the global corporate landscape. Organizations are experiencing a shift in mindset, moving from a sole focus on profit margins to a broader commitment to sustainability (Unruh et al., 2016). This shift shows a growing recognition of the need to adopt circular economy principles, which emphasize the conservation of resources, waste minimization, and environmental responsibility (Gupta et al., 2019).

One significant gap in achieving sustainability is the data gap, which has led to recent efforts such as the Digital Product Passport (DPP) as a policy instrument by the European Union to address this issue and implement circular economy principles (Walden et al., 2021). The DPP is a mechanism that supports data collection about a product's lifecycle, including its origin, materials, manufacturing processes, and end-of-use recovery, and further facilitates the exchange of information between different entities in value chains (World Business Council for Sustainable Development, 2023). To facilitate collaboration to obtain and store data, enterprises could use the capabilities of advanced tools such as blockchain and enterprise analytics.

Blockchain is among technologies that can facilitate circular economy by providing transparency, traceability, and accountability throughout the value chain (Centobelli et al., 2022). Blockchain provides an immutable ledger that tracks the lifecycle of products and promotes trust and collaboration among stakeholders. Equipping blockchain with enterprise analytics further facilitates real-time monitoring and data-driven decision-making and helps enterprises optimize processes for greater sustainability. Despite the significance of both blockchain and enterprise analytics, the combination of these two technologies in advancing circular economy is not well studied.

[1] Andrew F. Brimmer, College of Business and Information Sciences, Tuskegee University, Auburn, AL, 36088, USA.

[2,3] Environmental Engineering Sciences, University of Florida, Gainesville, FL, 32606, USA. Email: haoyuliao@ufl.edu[2]; sarabehdad@ufl.edu[3]

* Corresponding author: besmaeilian@tuskegee.edu

This book chapter aims to explore how the combination of enterprise analytics principles with blockchain technology facilitates achieving a circular economy. Our discussion is around three core concepts: (1) new sources of data, (2) innovative data collection technologies, and (3) advanced analytical approaches. Blockchain can be considered a modern product lifecycle management (PLM) system for collecting lifecycle data and strengthening the transparency of product lifespan (X. L. Liu et al., 2020).

Blockchain can be viewed as a modern PLM system as it provides a decentralized and immutable ledger that records every stage of a product's lifecycle. This transparent record-keeping in implementing DPPs improves visibility and accountability. Furthermore, blockchain's ability to facilitate secure data sharing among various parties supports practical collaboration to make informed decisions on sustainability initiatives.

The foundation of informed decision-making lies in data. While there are data gaps, on the other hand, enterprises have access to a wealth of previously untapped data sources (Günther et al., 2017). Modern organizations are teeming with cutting-edge data collection technologies. Drones, satellite imaging, and geospatial data provide a complete view of the entire product lifecycle, from raw material extraction to end-of-life disposal. Moreover, data analytics empowers organizations to extract valuable information from the data and move toward proactive decision-making (Jabbour et al., 2019).

Through discussing two case studies, we will show how product lifecycle data serves as the basis for making data-driven decisions that extend product lifecycles, reduce waste, and mitigate environmental impact. The first case study shows the importance of product serialization and how the data collected during the usage phase of laptop batteries can help analyze user behavior in charging and discharging batteries and predict the future reusability of each device based on its usage profile. The second case study discusses the role of product repair and maintenance data in predicting product failure, optimizing future purchase decisions, and proper workforce planning in repair shops. These case studies provide evidence of product lifecycle applications in diverse circular economy goals. For advancing sustainability through the implementation of product lifecycle engineering, it is important to collect product lifecycle data using distributed platforms such as blockchain, and it is equally important to utilize this data in a timely and practical manner.

This book chapter discusses the intersection of enterprise analytics and blockchain in advancing the circular economy and also explores how advancements in enterprise analytics mitigate some of the existing limitations of blockchain technology, particularly concerning data verification and validation, often referred to as the "last mile problem" (Gopalakrishnan and Behdad, 2019).

This study contributes to the literature by proposing a framework that integrates blockchain with enterprise analytics to support sustainable product lifecycle management. While existing studies have already discussed the use of blockchain for product lifecycle engineering and sustainability practices, the emphasis of this chapter is on the importance of product serialization and tracking of individual products, and most importantly, the need for analyzing such data.

The remainder of this chapter is organized as follows. Section 2 introduces three key developments in enterprise analytics. Section 3 discusses the importance of product lifecycle data for implementing circular economy principles and reviews two case studies to illustrate the impact of data on predicting product future reusability potential and extending product lifecycle. Section 4 describes how blockchain can help with product tracking and the way enterprise analytics can complement blockchain capabilities. Section 5 reviews some of the current limitations of blockchain on data verification and how enterprise analytics can alleviate some of these challenges, and finally, Section 6 concludes the chapter.

2. Enterprise Analytics

The new generation of enterprise analytics is characterized by several key developments as summarized in Figure 10.1, including new sources of data, new technologis, and emerging data analytics (T.H. Davenport, 2013), where each has significant implications for empowering blockchain technology. This section describes several recent advances in each area.

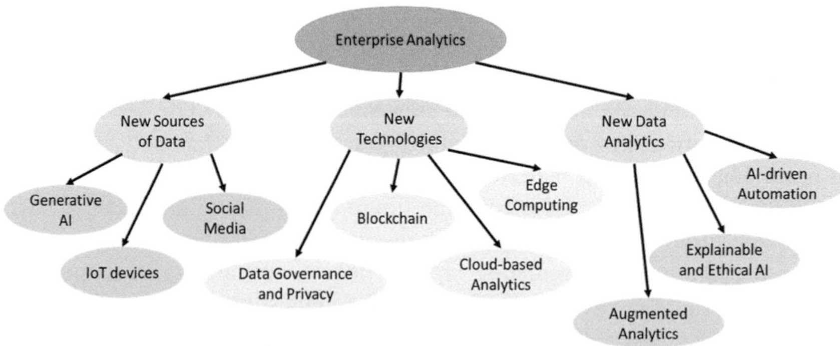

Fig. 10.1 The three key developments in enterprise analytics: (1) new data sources, (2) new technologies, and (3) new data analytics

2.1 New Sources of Data

The concept of big data has been emerging in recent years, with enterprises expanding the volume, velocity, and variety of their data collection efforts (Ohlhorst, 2012). The big data collected comes from several primary sources of data, including social media, machine data, and transactional data. Machine data indicates any data generated from industrial equipment, sensors, smart meters, IoT devices, and cameras, to name a few. The transactional data refers to all daily transactional data taking place online or offline such as payments, receipts, invoices, and operational records. Besides these resources, recently AI and ML methods are considered as new sources for data generation as AI models often generate new insights with the potential to be used as new streams of data for other decision-making processes.

Traditional structured data is now complemented by unstructured data from social media, IoT devices, and multimedia content (Blazquez and Domenech, 2018). Companies utilize these diverse data sources for improved decision-making. Data analytics methods such as natural language processing and sentiment analysis aid with the extraction of new data from text data. These new data sources provide the opportunity for organizations to increase their sustainable practices. The new sources of data help with improving circular economy performance and evaluating sustainable strategies. For example, smart sensors on farm equipment can optimize irrigation, reduce water wastage, and stimulate sustainable farming (Javaid et al., 2022). Besides, organizations can monitor public opinion towards their sustainable strategies by developing measurable indexes (Lee and Kim, 2021). For example, a company can track the public's response to its environmental initiatives on social media and receive public feedback (Cho et al., 2017). Table 10.1 provides examples of new sources of data available to organizations.

Table 10.1 Examples of sources of data available to enterprises

Data Source	*Description*
IoT Devices	Sensor data from Internet of Things (IoT) devices.
Social Media	User-generated content from social media platforms.
Machine-generated Data	Data generated by AI and machine learning algorithms.
Blockchain	Immutable transaction data from distributed ledgers.
Geospatial Data	Location-based information from GPS and satellites.

2.2 New Technologies

Enterprise analytics includes a range of cutting-edge technologies that empower organizations to employ the power of data for informed decision-making. Several examples include technologies for Data Governance and Privacy (Viljoen, 2021), Cloud-Based Analytics technologies (Khan et al., 2013), Blockchain (Swan, 2015), and Edge Computing (Shi et al., 2016).

Data governance is the management of data availability, usability, consistency, and security within an organization (Ladley, 2019). Privacy includes practices that protect sensitive data from unauthorized access. In enterprise analytics, data governance and privacy technologies include the implementation of policies, procedures, and tools to handle data securely. This includes data classification, access controls, and auditing mechanisms to track data usage. Recent advances in data governance and privacy include the development of automated tools for data classification, privacy impact assessments, and compliance monitoring (Janssen et al., 2020). Technologies such as artificial intelligence are used to detect anomalies and breaches in real-time, and blockchain technology is applied for immutable data records (Chithanuru and Ramaiah, 2023).

Cloud-based analytics includes running data analytics and processing tasks on cloud infrastructure rather than on-premises servers. It provides scalability and flexibility as organizations can access data from anywhere with an internet connection. Moreover, it helps businesses store and process large volumes of data while reducing the need for extensive IT infrastructure. Recent advancements in cloud-based analytics include serverless computing, multi-cloud, and hybrid cloud approaches that provide redundancy (Zhao et al., 2022).

Blockchain is also considered among enterprise analytics technologies as it serves as a platform for maintaining big data (Vo et al., 2018). Blockchain is a distributed ledger technology that records transactions within a network of computers and creates a transparent, tamper-proof, and secure record of data. Recent advances in blockchain include privacy-focused blockchains that create an environment for confidential transactions while still benefiting from blockchain's security and transparency features.

Finally, edge computing processes data closer to the data source, rather than relying on centralized cloud servers (Pansara, 2022). Edge computing provides the opportunity for real-time analysis of data generated by IoT devices, sensors, and other endpoints, and reduces latency and bandwidth usage. Edge analytics platforms are becoming more advanced as they provide data preprocessing, predictive maintenance, and anomaly detection tools. Data governance, privacy practices, cloud-based analytics, and edge computing technologies complement the security and privacy of blockchain networks. These advancements make it possible to integrate blockchain into a broader range of sustainability applications.

2.3 New Data Analytics Approaches

New data analytics approaches in enterprise analytics represent innovative approaches to extracting value from data. These approaches go beyond traditional descriptive analytics and include predictive and prescriptive analytics (Lepenioti et al., 2020). Predictive analytics uses historical and real-time data to forecast future trends and helps organizations make proactive decisions (McCarthy et al., 2022). For example, predictive analytics can help businesses anticipate customer demands, optimize inventory, and mitigate risks (Choi et al., 2018). Prescriptive analytics takes it a step further by both predicting outcomes and providing actionable recommendations to achieve desired results (Frazzetto et al., 2019). These approaches use advanced algorithms, machine learning models, and optimization strategies to guide decision-making. Together, predictive and prescriptive analytics help organizations optimize operations and drive strategic initiatives. Table 10.2 provides a summary of existing data analytics approaches along with common methods used in each category.

Table 10.2 Overview of data analytics approaches and key methods

Analytics Technique	*Description*	*Sample Methods*	*Sample Applications*
Descriptive Analytics	Historical data analysis to understand past events.	Data summary, visualization, reporting.	Tracking energy consumption patterns in buildings
Predictive Analytics	Utilizes historical and real-time data to forecast future trends.	Forecasting, data modeling, and machine learning.	Assessing future carbon emissions.
Prescriptive Analytics	Predicts outcomes and provides actionable recommendations.	Advanced algorithms, optimization strategies, and foundation models	Recommending sustainable supply chain practices

A wide range of data analytics techniques has been developed. Examples include AI-driven automation, data visualization tools, augmented analytics, foundation models, and explainable AI. AI-driven automation has revolutionized data processing and automated tasks from data preparation to modeling. Furthermore, the development of advanced visualization tools provides a more interactive and intuitive approach to data interpretation for a broader audience. Augmented analytics or smart data discovery is a new generation of enterprise analytics that reduces the reliance on human judgment and expertise by automatically extracting data patterns (T. Davenport and Fitts, 2021). Augmented analytics platforms, empowered by AI, assist users in knowledge discovery and recommendations. Foundation models, which undertake extensive pre-training on large-scale data, provide a flexible foundation for many downstream tasks, such as natural language processing, computer vision, and predictive analytics, with significant performance gains and increased flexibility (Bommasani et al., 2021). The ethical use of AI and the rise of Explainable AI (XAI) reinforce trust and accountability. These advances collectively help organizations leverage data for more informed decision-making.

3. The Value of Product Lifecycle Data for Sustainable Practices

Collecting product lifecycle data through product lifecycle management systems is important for achieving circular economy objectives in modern business operations (Ren et al., 2019). The lifecycle data helps organizations make informed decisions that minimize environmental impact and improve customer satisfaction. One significant benefit of collecting product lifecycle data is the ability to personalize the customer experience and product-service systems (Hribernik et al., 2017). Businesses can tailor their services to align with individual customer needs by analyzing the product's lifecycle. Moreover, product lifecycle data helps with optimizing the entire value chain, from raw material sourcing to manufacturing, distribution, and responsible disposal by facilitating collaboration and responsibility sharing among various stakeholders (Chiang and Trappey, 2007; Jacobs and Subramanian, 2012).

Accurate product lifecycle data is necessary for forecasting a product's lifespan. This data assists organizations to predict when a product will reach the end of its useful life and plan for responsible disposal, recycling, or repurposing. Monitoring consumption habits and implementing predictive maintenance is another usage for product lifecycle data. Figure 10.2 summarizes examples of the benefits of product lifecycle data for achieving a circular economy along with current challenges and solutions for addressing some of those challenges.

Benefits of Product Lifecycle Data	Challenges of Product Lifecycle Data	Solutions
• Personalizing customer experience and customized services • Optimizing value chain • Forecasting product lifespan • Monitoring consumption habits for predictive maintenance	• Data fragmentation and limited access • Data acquisition and management • Lack of actionable information for decision-making • Insufficient data collection infrastructure	• Emerging data sources • Innovative data collection methods • Advanced data analytics

Fig. 10.2 Examples of benefits, challenges, and solutions in using product lifecycle data

To clarify the application of product lifecycle data, we will briefly discuss two case studies: (1) we explain how individual battery lifecycle data can be used to predict usage behavior and assess the future reusability of individual batteries, and (2) we describe how the repair and maintenance log data of products can aid in decision making such as predicting the demand for repair, product lifespan, and the time required for repair.

We acknowledge that the data for these case studies were manually connected, and not sourced from any PLM or distributed systems. Although these case studies do not directly include blockchain, they show the potential benefits of gathering accurate lifecycle data and product serialization. Manual collection of product lifecycle data is not feasible, as current practices suffer from issues of data reliability and accountability. The challenges of collecting data throughout a product's entire lifespan include the lack of consistent data collection standards, fragmented data ownership, and limited stakeholder collaboration. Blockchain as an immutable and transparent system can overcome some of these issues since it distributes the data in a network of stakeholders, eliminates the need for a central authority, and reduces the risk of data loss or manipulation. Further, blockchain can be integrated with IoT devices and sensors to automatically capture and record data at each stage of the product lifecycle, so it reduces the reliance on manual data entry (C. H. Liu et al., 2018).

3.1 *Case Study 1: User Consumption Habits and Forecasting Product Reusability*

Smart asset management provides users the benefit of monitoring a product during its entire lifespan and the possibility of prolonging the product lifecycle, which is a circular economy cornerstone. This case describes how collecting individual device data helps remanufacturers monitor each unique device's performance and personalize their service based on the needs of individual users. To clarify the concept, we will review a case of monitoring lithium-ion laptop batteries. The data presented in this case study has been collected over three years on 500 laptop batteries used by students in an all-girl high school in Illinois, USA as part of a prior study published in (Sabbaghi et al., 2015b, 2015a).

The dataset includes battery weight, serial number, type, voltage, used cycles, full charge capacity, battery age, and the class of students. To explain the dataset, Figure 10.3 (a) shows the boxplot of batteries' current capacity, and Figure 10.3 (b) illustrates the voltage versus the current capacity of batteries of different ages. The variation in the dataset reveals the importance of collecting data on individual devices. Users have different charging and discharging behaviors that influence the number of used cycles and the lifecycle of individual batteries.

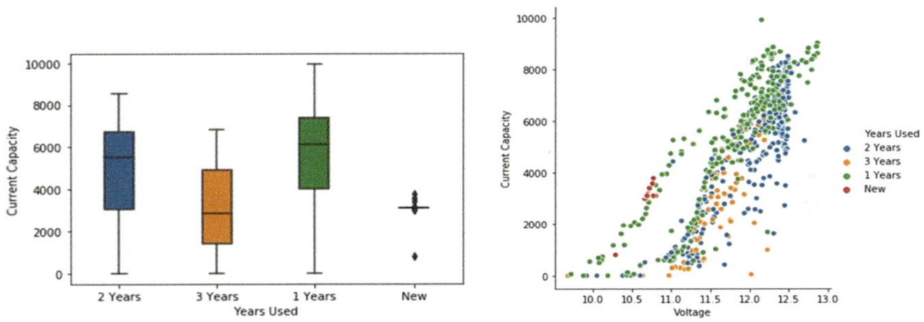

(a) (b)

Fig. 10.3 (a) The current capacity, and (b) voltage versus current capacity of batteries with 1 year, 2 years, and 3 years of usage.

The product lifecycle data helps remanufacturers customize their end-of-use decision on whether to reuse or retire batteries based on the number of remaining cycles and the potential for future device reusability. Moreover, the products offered in the second-hand market can be appropriately priced based on their potential for reuse and remanufacturing. For example, Figure 10.4 shows the distribution of the number of used cycles for the 500 batteries over three years. Consider a value of 300 used cycles as the threshold for degrading the battery. From the distribution of the used cycles and the area under the curve, we can calculate the likelihood that the battery has 200 remaining used cycles and the chance of future reusability. The authors have conducted a battery reusability analysis study on this

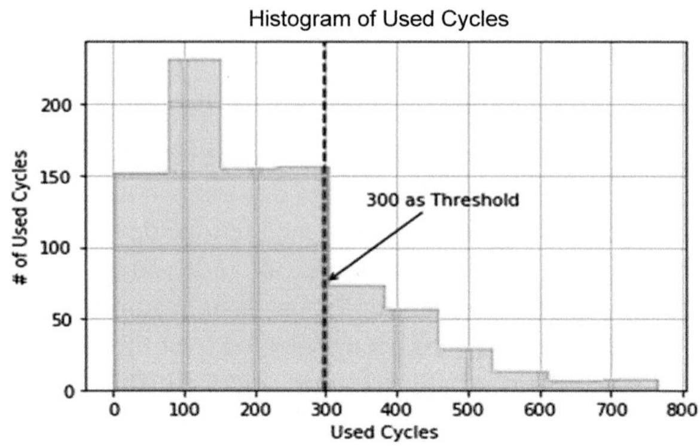

Fig. 10.4 The distribution of the number of used cycles

dataset. For more information, we refer readers to (Sabbaghi et al., 2015b, 2015a). Overall, the concept of Carfax used in the automobile industry can be employed for other smart devices where the data on the product's profile of usage, repair, and upgrade would give some information about the device's lifespan and future user experience. Moreover, the product lifecycle data could also help remanufacturers decide when would be a good time to offer upgrade, repair, and maintenance services for individual devices.

3.2 Case Study 2: Product Lifecycle Data for Optimizing Repair, Maintenance, and Upgrade Decisions

This case study discusses the repair and maintenance log data for three types of medical devices. It provides descriptive statistics on the frequency and nature of product failures, along with information such as mean time to failure, mean time to repair, and total operating time of these devices. The objective is to show the practical usage of this data in improving decision-making for healthcare providers when purchasing devices. In addition, it shows how repair shops can use this data to optimize human resource allocation and job scheduling in response to repair demand and specific repair requirements.

The data includes three rebranded medical devices: M1, M2, and P1, that serve the function of infusion pumps and multitherapy. These labels are assigned as per the initial letter of each device's name. Table 10.3 provides an overview of these medical devices. M1 has the highest quantity of individual units, while P1 has the most extended period of usage, which is approximately 14 years from 2004 to 2018.

These devices show a similar average failure count, ranging from 8 to 10, as listed in Table 10.3. Table 10.4 shows the failure causes requiring maintenance for these products. The dataset includes 12 different failure reasons, including battery-related issues, random failures, and instances where no problems are detected, among others. Among these reasons, the three main drivers of maintenance are battery-related concerns, random failures, and cases where no issues are identified.

Table 10.3 Basic information of three product categories.

Product Category	*Function Description*	*Number of maintenance records*	*Number of individual products*	*First and last record date*	*The average of Failure count*
M1	Infusion Pumps, Multitherapy	15903	1593	04-23-2004 / 06-24-2011	10
M2		4274	599	02-18-2004 / 10-10-2011	7
P1		6085	805	01-06-2004 / 01-11-2018	8

Table 10.4 The number of repair and maintenance records for each product for different reasons.

Maintenance Description	*M1*	*M2*	*P1*
Battery Related (BR)	5859	533	1887
Random Failure (RF)	5487	2229	826
No Problem Found (NPF)	2475	731	1251
Modification/Upgrade (MU)	571	87	812
Use Error (UE)	385	119	352
Physical Damage (PD)	240	111	640
Random Software Failure (RSF)	202	116	139
Accessory Problem (AP)	56	10	119
PM Related (PMR)	19	4	59
Tech Support (RS)	605	334	0
Environment Related (ER)	3	0	0
Normal Retire/Replacement (NR)	1	0	0

While M1 and P1 share battery-related problems as their primary maintenance concern, M2's primary maintenance need is attributed to random failures.

Figure 10.5 presents the maintenance records for each product annually. Figure 10.5 (a) illustrates the period from 2004 to 2007, during which each product shows a similar trend. All three products experienced their initial peak in maintenance cases in 2006. M1 follows a different trajectory compared to M2 and P1. After 2011, only a few P1 products were used while it seems M1 and M2 are retired based on the source of the dataset.

Figures 10.5 (b) and (c) underline the battery-related maintenance, which becomes major after 2007 for all three products. The random failures consistently show up as the primary maintenance reason every year for M1 and M2. However, a shift occurred after 2008, as battery-related issues became the primary maintenance concern for M1 and P1.

We could look at individual device's lifecycle data to personalize the needs of each device. Figure 10.6 presents a product lifecycle analysis of three individual

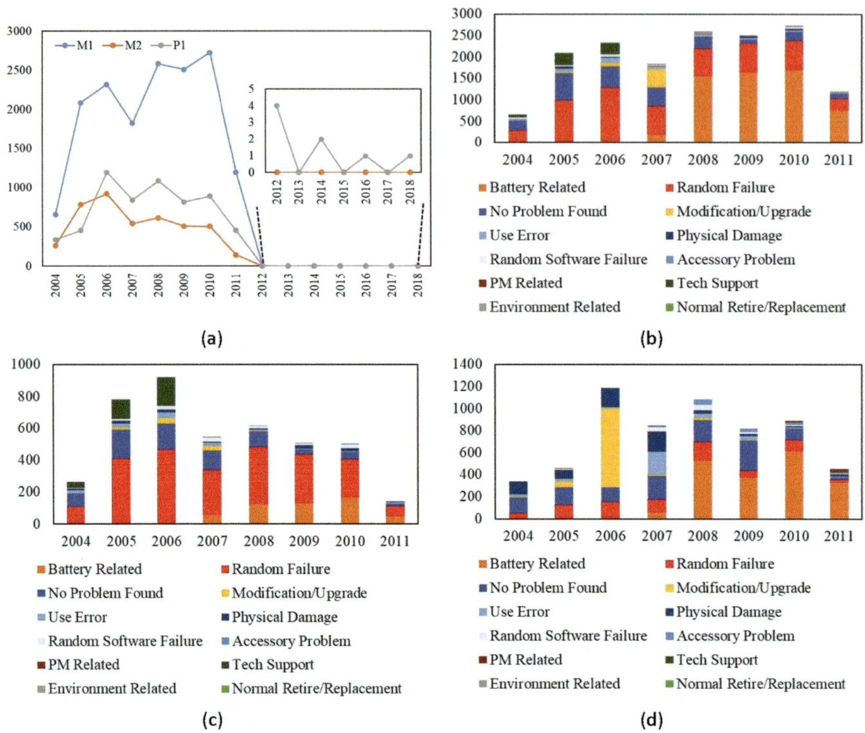

Fig. 10.5 Annual Maintenance Records: (a) All Failures, and Specific Reasons for (b) M1, (c) M2, and (d) P1

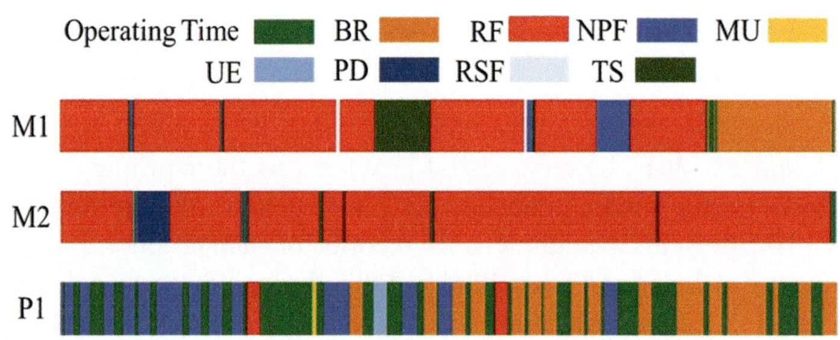

Fig. 10.6 Product Lifecycle Analysis for M1 (Product ID: 01011043), M2 (Product ID: 01001859), and P1 (Product ID: 91015346). Note: Operating time is measured in years, while all other units are in days. The abbreviation labels are described in Table 10.4.

devices: M1 (ID 01011043), M2 (ID 01001859), and P1 (ID 91015346). Table 10.5 complements this analysis by providing key metrics such as Mean Time Between Failure (MTBF), Mean Time to Repair (MTTR), Total Operating Time (OT), and Total Repairing Time (RT).

According to Figure 10.6, M1 and M2 share a common main reason for repair, namely random failure. On the other hand, P1 has two primary reasons for maintenance, characterized as "no problem found" and "battery-related." The product lifecycle evaluation also highlights that P1 shows comparatively less downtime than M1 and M2, which consequently reduces repair costs.

Access to detailed repair and maintenance data has benefits for both users and repair shops. For users, such information provides guidelines for when to repair or replace a product. It also extends the product's lifespan through proactive maintenance measures by reducing unexpected breakdowns and operational downtime. Users can make choices regarding which products to purchase, based on their historical maintenance performance. On the other hand, repair shops benefit from this data due to better resource planning and job scheduling with minimum downtime for users. The data also helps repair shops practically address common issues and improve their service quality.

Table 10.5 Information for Each Individual Device

Product	*ID*	*Failure count*	*MTTB (Days)*	*MTTR (Days)*	*Total OT (Days)*	*Total RT (Days)*
M1	1011043	34	67	9.3	2211	317
M2	1001859	24	88	12	2025	292
P1	91015346	38	72	0.3	2665	13

The above discussions were mainly around reporting the available data to show the capability of enterprise analytics to help organizations optimize their sustainability practices. The authors have conducted several analyses using machine learning techniques to predict repair demand, the type of failure, and the time to repair devices. For more information, we refer readers to Liao et al., 2022; Liao, Boregowda, et al., 2021; Liao, Cade, et al., 2021.

Although manual collection of product lifecycle data is beneficial, current manual practices are not practical and suffer from issues of data reliability, integrity, and accountability. Often, the data required by one stakeholder in a specific phase of the product lifecycle should be collected by another stakeholder during a different phase. For example, data needed at the end of a product's use should ideally be collected during the initial design and usage phases, with stakeholders incentivized to gather such information for remanufacturers and recyclers. Current policies, such as Extended Producer Responsibility (EPR) for e-waste, are not successful due to the lack of a reliable product tracking system. Once products are sold, OEMs have no control over end users, and it is uncertain when a product will be ready for recycling or end-of-usage discarding (Leclerc and Badami, 2020; Ahlers et al., 2021). In addition, there is a disconnect between business objectives and sustainability goals, where enterprises often ignore the cost-saving opportunities that product tracking and the consequent sustainability initiatives can bring (Kunz et al., 2018). To address these challenges, it is important to establish novel business models that incentivize and reward stakeholders, from OEMs to end users and recyclers, for their participation and collaboration in building a reliable data infrastructure.

4. Product Tracking with Blockchain and Enterprise Analytics

Product tracking assists with sustainable practices such as rewarding consumers for their green behavior (Kahya et al., 2021), personalizing end-of-use decisions for discarded devices based on their usage profiles, looking up repair and maintenance instructions (J. Stodt et al., 2019), certifying transparency of the use of toxic materials and material consumption (Petrović et al., 2022), tracing sources of materials, informing product designers about the impact of their designs on product end-of-use recovery fate (Ibarra et al., 2024), and providing visibility of formal and informal trades of products among regions to confirm compliance with regulations.

Product tracking is motivated by both economic benefits recognized by enterprises as well as sustainability-related policy mechanisms such as the EU's Digital Product Passport. Figure 10.7 shows how the DPP regulatory framework can be implemented by using the capabilities of blockchain and analytics.

The major objective of DPP is to support the sharing of key product-related information, essential for sustainability and circularity, among all relevant economic actors, and provide new business opportunities for them. It also aims to assist consumers in making sustainable decisions and allows authorities to verify compliance with legal obligations.

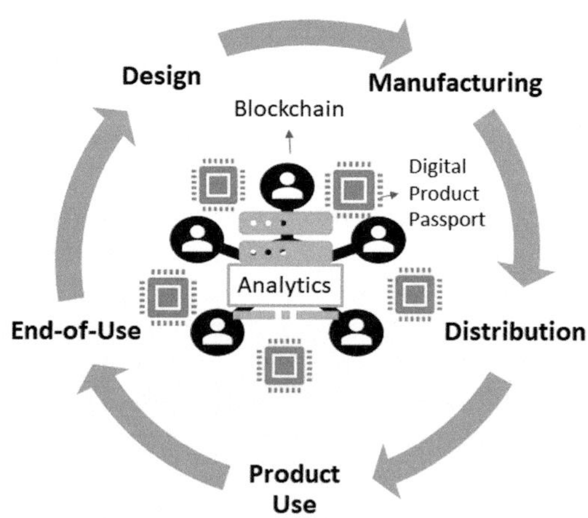

Fig. 10.7 Implementation of digital product passport regulatory framework through blockchain and data analytics

The implementation of the EU DPP encounters several challenges, which can be grouped into three categories: scope, technology, and data. Scope-related challenges include determining which industries and product groups should be prioritized, whether requirements should vary by company size, and the appropriate application level of DPPs, such as item, batch, or product model level. Technological challenges include decisions on how and by whom data should be stored, the choice of data carriers, and how access to the data should be managed and secured. Data-related challenges include defining what information will be included in the DPP and to what degree it should be standardized, as well as establishing who is responsible for collecting and updating the data and how its accuracy and quality will be verified (World Business Council for Sustainable Development, 2023).

Blockchain platforms help overcome some of the technological challenges of DPP related to data storage, data carrier choice, and access/security by providing decentralized, secure, and transparent data management solutions (F. Stodt et al., 2024). Enterprise analytics has the potential to address data-related challenges of DPP by advancing data collection, analysis, and verification processes.

Blockchain is known for its capability in recording and tracking product lifecycle data (Matenga and Mpofu, 2023). Recently, it has been recognized for circular economy principles due to characteristics such as transparency, traceability, and accountability (Centobelli et al., 2022). Blockchain is a decentralized, distributed ledger technology in which each user in the network has access to a copy of the ledger and the information is securely recorded on the ledger with some characteristics such as transparency, immutability, and cryptographic integrity. Therefore, for various stakeholders during the product lifespan who conventionally had some concerns about the security and privacy of their data, Blockchain can alleviate such concerns.

5. Current Limitations of Blockchain

Despite capabilities such as immutability and traceability, Blockchain technology still suffers from several limitations such as scalability challenges, high energy consumption, limited transaction throughput, potential privacy concerns, and lack of proper verification techniques known as the "last mile problem" (Esmaeilian et al., 2024; Y. Lu, 2019; Zheng et al., 2018).

The "last mile problem" is the challenge of confirming the accuracy and authenticity of data before it is recorded on the blockchain ledger as blockchain is immutable and tamper-proof. This challenge includes several complexities that demand careful consideration.

Several methods and solutions have been proposed to address this challenge. One solution is the use of Oracle services. An "oracle" points to a trusted external data source or service that provides real-world information to a blockchain network. Oracles act as intermediaries that verify and provide external data to the blockchain (Ezzat et al., 2022). Another method is the use of data verification protocols. These protocols include multiple steps of data verification before it is committed to the blockchain. Techniques such as multi-signature scheme (Uganya and Baskar, 2023) and consensus mechanisms within decentralized autonomous organizations (Wang et al., 2019) can be used to check that data is agreed upon by multiple parties before it is recorded. Also, adding an additional layer for auditing and verifying data before it is recorded on the blockchain is another solution. This can include third-party auditors or automated verification tools that cross-check data against predefined criteria (Chen et al., 2022). Moreover, smart contracts can be programmed to perform specific checks and validations on data before it is accepted. This includes logical checks, cross-referencing with other data sources, and compliance with certain standards (W. Lu et al., 2023). Further, the use of cryptographic techniques such as zero-knowledge proofs allow data providers to prove the validity of their data without revealing the data itself (Sun et al., 2021).

Addressing the last mile problem in data verification and validation is necessary for showing the trustworthiness of blockchain technology, particularly in use cases such as supply chain management and sustainable development applications.

Besides the above-mentioned methods, enterprise analytics provide solutions for addressing the existing data verification issues on blockchain networks. Using sensors, IoT, social media data, and advanced machine learning and AI techniques can address some of the challenges associated with the "last mile problem". To name several examples, real-time data from *sensors and IoT devices* can provide continuous inputs to the blockchain. These devices offer on-the-ground information and improve the timeliness of data. Also, *machine learning models* can be used to verify incoming data. AI algorithms can detect anomalies or discrepancies and identify potential inaccuracies. Furthermore, machine learning and AI can be integrated into smart contracts to provide autonomous decision-making and reduce the need for human intervention. Also, machine learning algorithms can optimize consensus mechanisms, and facilitate faster agreement among nodes in case of verifying the authenticity of incoming data. Moreover, combining data from various sources, including sensors, IoT, and social media, can increase the reliability of information. *Advanced fusion techniques* can cross-verify and validate data points and reduce the risk of errors. Data fusion is an integration of new sources of data and the use of data analytics techniques to cross-validate the data.

6. Conclusion

This chapter explores the intersection between enterprise analytics and blockchain for advancing circular economy goals, with an emphasis on facilitating data collection during the product lifecycle. Motivated by recent policy instruments such as the EU's digital product passports, this chapter discussed the importance of product lifecycle data for sustainable practices through two case studies and explained the potential of blockchain to collect and manage this data. It also described how enterprise analytics can complement sustainability efforts by providing new sources of data, advanced technologies, and sophisticated data analytics. These developments help organizations gather diverse data streams, optimize operations, and implement sustainable decision-making.

As organizations increasingly turn to blockchain for data integrity, the "last mile problem" has become a critical challenge, that requires solutions for accurate data verification. Enterprise analytics can mitigate this issue by providing real-time data integration, machine learning verification, and data fusion techniques.

There are several limitations to the current study that should be acknowledged. The case studies presented are limited in scope and may not be representative of all industries or circular economy practices. Also, the data collected in the case studies were manually gathered. While the chapter discusses the theoretical framework and potential benefits of blockchain and enterprise analytics, these concepts should be validated through future practical implementation. Moreover, this study mainly describes the technological aspects of sustainability, namely blockchain and enterprise analytics. However, it does not investigate the role of novel business models and value-driven strategies based on sustainability. For a truly sustainable practice, businesses should rethink their commercial structures and innovate to create new opportunities that solve social problems and achieve sustainability.

The integration of enterprise analytics and blockchain technology to achieve circular economy goals opens an outlook of unexplored potential. Future work includes investigating the ethical implications of data collection and sharing, exploring the scalability challenges of blockchain networks and the development of new data sources and analytics approaches, as well as successful implementation of use cases in practice that show the success of organizations toward a more sustainable and customer-centric future.

Acknowledgment

The US National Science Foundation has provided financial support for the conduct of the research under Grant # 2324950 for Hao-yu Liao and Sara Behdad. Any opinions, findings, conclusions, or recommendations expressed in this material are those of the authors and do not necessarily reflect the views of the National Science Foundation.

References

Ahlers, J., Hemkhaus, M., Hibler, S. and Hannak, J. (2021). *Analysis of Extended Producer Responsibility Schemes:* Assessing the performance of selected schemes in European and EU countries with a focus on WEEE, waste packaging, and waste batteries. adelphi consult GmbH. Retrieved from https://erp-recycling.org/wp-content/uploads/2021/07/adelphi_study_Analysis_of_EPR_Schemes_July_2021.pdf.

Blazquez, D. and Domenech, J. (2018). Big Data sources and methods for social and economic analyses. *Technological Forecasting and Social Change*, *130*, 99–13.

Bommasani, R., Hudson, D.A., Adeli, E., Altman, R., Arora, S., von Arx, S., Bernstein, M.S., Bohg, J., Bosselut, A. and Brunskill, E. (2021). On the opportunities and risks of foundation models. *ArXiv Preprint ArXiv:2108.07258*.

Centobelli, P., Cerchione, R., Del Vecchio, P., Oropallo, E. and Secundo, G. (2022). Blockchain technology for bridging trust, traceability and transparency in circular supply chain. *Information & Management*, *59*(7), 103508.

Chen, L., Fu, Q., Mu, Y., Zeng, L., Rezaeibagha, F. and Hwang, M.-S. (2022). Blockchain-based random auditor committee for integrity verification. *Future Generation Computer Systems*, *131*, 183–93.

Chiang, T.-A. and Trappey, A.J.C. (2007). Development of value chain collaborative model for product lifecycle management and its LCD industry adoption. *International Journal of Production Economics*, *109*(1–2), 90–104.

Chithanuru, V. and Ramaiah, M. (2023). An anomaly detection on blockchain infrastructure using artificial intelligence techniques: Challenges and future directions–A review. *Concurrency and Computation: Practice and Experience*, *35*(22), e7724.

Cho, M., Furey, L.D. and Mohr, T. (2017). Communicating corporate social responsibility on social media: Strategies, stakeholders, and public engagement on corporate Facebook. *Business and Professional Communication Quarterly*, *80*(1), 52–69.

Choi, T., Wallace, S.W. and Wang, Y. (2018). Big data analytics in operations management. *Production and Operations Management*, *27*(10), 1868–83.

Davenport, T. and Fitts, J. (2021). AI Can Help Companies Tap New Sources of Data for Analytics. *Harvard Business Review*. https://hbr.org/2021/03/ai-can-help-companies-tap-new-sources-of-data-for-analytics

Davenport, T.H. (2013). *Enterprise analytics: Optimize performance, process, and decisions through big data*. Pearson Education.

Esmaeilian, B., Jamison, M., Sarkis, J. and Behdad, S. (2024). Designing Future Sustainable Cryptocurrencies: Principles and Expectations. In *Blockchain and Smart-Contract Technologies for Innovative Applications* (pp. 59–87). Springer.

Ezzat, S.K., Saleh, Y.N.M. and Abdel-Hamid, A.A. (2022). Blockchain oracles: State-of-the-art and research directions. *IEEE Access*, *10*, 67551–67572.

Frazzetto, D., Nielsen, T.D., Pedersen, T.B. and Šikšnys, L. (2019). Prescriptive analytics: a survey of emerging trends and technologies. *The VLDB Journal*, *28*, 575–95.

Gopalakrishnan, P.K. and Behdad, S. (2019). A Conceptual Framework for Using Videogrammetry in Blockchain Platforms for Food Supply Chain Traceability. *ASME 2019 International Design Engineering Technical Conferences and Computers and Information in Engineering Conference*.

Günther, W.A., Mehrizi, M.H.R., Huysman, M. and Feldberg, F. (2017). Debating big data: A literature review on realizing value from big data. *The Journal of Strategic Information Systems*, *26*(3), 191–209.

Gupta, S., Chen, H., Hazen, B.T., Kaur, S. and Gonzalez, E.D.R.S. (2019). Circular economy and big data analytics: A stakeholder perspective. *Technological Forecasting and Social Change*, *144*, 466–74.

Hribernik, K., Franke, M., Klein, P., Thoben, K.-D. and Coscia, E. (2017). Towards a platform for integrating product usage information into innovative product-service design. *2017 International Conference on Engineering, Technology and Innovation (ICE/ITMC)*, 1407–13.

Ibarra, D.S., Li, F., Zhu, J. and Chen, J. (2024). Blockchain application to the processes in material design, production, distribution, and disposal: A survey. *Journal of Industrial Information Integration*, 100638.

Jabbour, C.J.C., de Sousa Jabbour, A.B.L., Sarkis, J. and Godinho Filho, M. (2019). Unlocking the circular economy through new business models based on large-scale data: an integrative framework and research agenda. *Technological Forecasting and Social Change*, *144*, 546–52.

Jacobs, B.W. and Subramanian, R. (2012). Sharing responsibility for product recovery across the supply chain. *Production and Operations Management*, *21*(1), 85–100.

Janssen, M., Brous, P., Estevez, E., Barbosa, L.S. and Janowski, T. (2020). Data governance: Organizing data for trustworthy Artificial Intelligence. *Government Information Quarterly*, *37*(3), 101493.

Javaid, M., Haleem, A., Singh, R. P., & Suman, R. (2022). Enhancing smart farming through the applications of Agriculture 4.0 technologies. *International Journal of Intelligent Networks*, *3*, 150–64.

Kahya, A., Avyukt, A., Ramachandran, G.S. and Krishnamachari, B. (2021). Blockchain-enabled Personalized Incentives for Sustainable Behavior in Smart Cities. *2021 International Conference on Computer Communications and Networks (ICCCN)*, 1–6.

Khan, Z., Anjum, A. and Kiani, S.L. (2013). Cloud based big data analytics for smart future cities. *2013 IEEE/ACM 6th International Conference on Utility and Cloud Computing*, 381–86.

Kunz, N., Mayers, K. and Van Wassenhove, L.N. (2018). Stakeholder views on extended producer responsibility and the circular economy. *California Management Review*, *60*(3), 45–70.

Ladley, J. (2019). *Data governance: How to design, deploy, and sustain an effective data governance program*. Academic Press.

Leclerc, S.H. and Badami, M.G. (2020). Extended producer responsibility for E-waste management: Policy drivers and challenges. *Journal of Cleaner Production*, *251*, 119657.

Lee, R. and Kim, J. (2021). Developing a social index for measuring the public opinion regarding the attainment of sustainable development goals. *Social Indicators Research*, *156*, 201–21.

Lepenioti, K., Bousdekis, A., Apostolou, D. and Mentzas, G. (2020). Prescriptive analytics: Literature review and research challenges. *International Journal of Information Management*, *50*, 57–70.

Liao, H., Boregowda, K., Cade, W. and Behdad, S. (2021). Machine Learning to Predict Medical Devices Repair and Maintenance Needs. *International Design Engineering Technical Conferences and Computers and Information in Engineering Conference*, *85413*, V005T05A012.

Liao, H., Cade, W. and Behdad, S. (2021). Markov chain optimization of repair and replacement decisions of medical equipment. *Resources, Conservation and Recycling*, *171*, 105609.

Liao, H., Cade, W. and Behdad, S. (2022). Forecasting repair and maintenance services of medical devices using support vector machine. *Journal of Manufacturing Science and Engineering*, *144*(3), 31005.

Liu, C. H., Lin, Q. and Wen, S. (2018). Blockchain-enabled data collection and sharing for industrial IoT with deep reinforcement learning. *IEEE Transactions on Industrial Informatics*, *15*(6), 3516–26.

Liu, X. L., Wang, W.M., Guo, H., Barenji, A.V., Li, Z. and Huang, G.Q. (2020). Industrial blockchain based framework for product lifecycle management in industry 4.0. *Robotics and Computer-Integrated Manufacturing*, *63*, 101897.

Lu, W., Lou, J. and Wu, L. (2023). Combining smart construction objects–enabled blockchain oracles and signature techniques to ensure information authentication and integrity in construction. *Journal of Computing in Civil Engineering*, *37*(6), 4023031.

Lu, Y. (2019). The blockchain: State-of-the-art and research challenges. *Journal of Industrial Information Integration*, *15*, 80–90.

Matenga, A.E. and Mpofu, K. (2023). Blockchain-based product lifecycle management using supply chain management for railcar remanufacturing. *Procedia CIRP*, *116*, 486–91.

McCarthy, R.V, McCarthy, M.M., Ceccucci, W., Halawi, L., McCarthy, R.V, McCarthy, M.M., Ceccucci, W. and Halawi, L. (2022). *Applying predictive analytics*. Springer.

Ohlhorst, F.J. (2012). *Big data analytics: turning big data into big money* (Vol. 65). John Wiley & Sons.

Pansara, R.R. (2022). Edge Computing in Master Data Management: Enhancing Data Processing at the Source. *International Transactions in Artificial Intelligence*, *6*(6), 1–11.

Petrović, E K., Brent, A.C., Iorns Magallanes, C., Hamer, L. and van Eijck, D. (2022). Blockchain for supply chain ledgers: tracking toxicity information of construction materials. In *Blockchain for Construction* (pp. 89–111). Springer.

Ren, S., Zhang, Y., Liu, Y., Sakao, T., Huisingh, D. and Almeida, C.M.V.B. (2019). A comprehensive review of big data analytics throughout product lifecycle to support sustainable smart manufacturing: A framework, challenges and future research directions. *Journal of Cleaner Production*, *210*, 1343–65.

Sabbaghi, M., Esmaeilian, B., Raihanian Mashhadi, A., Cade, W. and Behdad, S. (2015a). Assessment of Products Future Reusability Based on Consumers Usage Behavior: Implications for Lithium-ion Laptop Batteries. *ASME 2015 International Design Engineering Technical Conferences & Computers and Information in Engineering Conference (IDETC/CIE 2015)*.

Sabbaghi, M., Esmaeilian, B., Raihanian Mashhadi, A., Cade, W. and Behdad, S. (2015b). Reusability assessment of lithium-ion laptop batteries based on consumers actual usage behavior. *Journal of Mechanical Design*, *137*(12), 124501.

Shi, W., Cao, J., Zhang, Q., Li, Y. and Xu, L. (2016). Edge computing: Vision and challenges. *IEEE Internet of Things Journal, 3*(5), 637–46.

Stodt, F., Maisch, N., Ruf, P., Lechler, A., Riedel, O. and Reich, C. (2024). *Collaborative Smart Production Supply Chains with Blockchain Based Digital Product Passports.*

Stodt, J., Jastremskoj, E., Reich, C., Welte, D. and Sikora, A. (2019). Formal description of use cases for industry 4.0 maintenance processes using blockchain technology. *2019 10th IEEE International Conference on Intelligent Data Acquisition and Advanced Computing Systems: Technology and Applications (IDAACS), 2*, 1136–41.

Sun, X., Yu, F. R., Zhang, P., Sun, Z., Xie, W. and Peng, X. (2021). A survey on zero-knowledge proof in blockchain. *IEEE Network, 35*(4), 198–205.

Swan, M. (2015). *Blockchain: Blueprint for a new economy.* O'Reilly Media, Inc.

Uganya, G. and Baskar, R. (2023). Modified Elliptic Curve Cryptography Multi-Signature Scheme to Enhance Security in Cryptocurrency. *Computer Systems Science & Engineering, 45*(1).

Unruh, G., Kiron, D., Kruschwitz, N., Reeves, M., Rubel, H. and Zum Felde, A.M. (2016). Investing for a sustainable future: Investors care more about sustainability than many executives believe. *MIT Sloan Management Review, 57*(4).

Viljoen, S. (2021). A relational theory of data governance. *Yale LJ, 131*, 573.

Vo, H.T., Mohania, M., Verma, D. and Mehedy, L. (2018). Blockchain-powered big data analytics platform. *Big Data Analytics: 6th International Conference, BDA 2018, Warangal, India, December 18–21, 2018, Proceedings 6*, 15–32.

Walden, J., Steinbrecher, A. and Marinkovic, M. (2021). Digital product passports as enabler of the circular economy. *Chemie Ingenieur Technik, 93*(11), 1717–27.

Wang, S., Ding, W., Li, J., Yuan, Y., Ouyang, L. and Wang, F.-Y. (2019). Decentralized autonomous organizations: Concept, model, and applications. *IEEE Transactions on Computational Social Systems, 6*(5), 870–78.

World Business Council for Sustainable Development. (2023). *The EU Digital Product Passport shapes the future of value chains: What it is and how to prepare.*

Zhao, H., Benomar, Z., Pfandzelter, T. and Georgantas, N. (2022). Supporting Multi-Cloud in Serverless Computing. *2022 IEEE/ACM 15th International Conference on Utility and Cloud Computing (UCC)*, 285–90.

Zheng, Z., Xie, S., Dai, H.-N., Chen, X. and Wang, H. (2018). Blockchain challenges and opportunities: A survey. *International Journal of Web and Grid Services, 14*(4), 352–75.

11 Blockchain's Transformative Potential for Circular Economy Laws and Policies

Katrien Steenmans[1], Ine Steenmans[2] and Phillip Taylor[3]

1. Introduction

Planetary boundaries define the levels at which humans and other species can safely operate without risking major global human-induced environmental change (Rockström et al., 2009; Richardson et al., 2023). Most planetary boundaries are reached by the deposit of waste back into the environment (Richardson and Bai, 2023): climate change, novel entities (i.e. anthropogenic introductions such as synthetic chemicals and substances, genetically modified organisms, and anthropogenically mobilised radioactive materials), stratospheric ozone depletion, atmospheric aerosol loading, ocean acidification, and biochemical flows (Rockström et al., 2009; Richardson et al., 2023).[1] At the time of writing, three of these boundaries have already been breached (Rockström et al., 2009; Richardson et al., 2023). In part, this stems from our high resource consumption levels—1.7 Earths are needed to support our current rates of natural resource consumption (Global Footprint Network, 2018). The resulting concomitant resource and waste crises have therefore prompted the reactive creation of a plethora of laws and policies aiming to achieve sustainable development and resource efficiency. Part of those aims include a global transition towards circular economies, which is our focus in this chapter.

The circular economy moves from a linear take-make-dispose economy to a system in which resource use and waste production are minimised and, if resources are used and wastes are generated, they are 'looped back' through reuse, recycling, and recovery. The introduction and operationalisation of circular economies have been largely dependent on the development of voluntary schemes and industry self-regulation, with only recently a growing number of laws and policies promoting

[1] University of Copenhagen, Denmark.
[2] University College London, UK.
[3] University of Warwick, UK.
Email: ine.steenmans@ucl.ac.uk[2]; phillip.taylor@warwick.ac.uk[3]
* Corresponding author: katrien.steenmans@jur.ku.dk

[1] The other three planetary boundaries are: freshwater change, land-system change, and biosphere integrity. These three boundaries have also been breached (Richardson et al., 2023).

the concept explicitly (Lesniewska and Steenmans, 2023). These legal and policy instruments have several limitations. There is, for example, no universal legal understanding of the concept. The limited definitions within laws and policies generally centre around: (1) minimising resource consumption and preventing the generation of wastes, and (2) reuse and recycling of wastes and resources within production and consumption processes (Lesniewska and Steenmans, 2023). There is also recognition that circular economies need to incorporate social dimensions, including justice and equity (Friant et al., 2020; Schröder, 2020; Lesniewska and Steenmans, 2023), mirroring the trend in other interventions that increasingly emphasise the need for a just transition (e.g. climate change and shifting away from fossil fuels; see McCauley and Heffron, 2018; Wang and Lo, 2021). Such circular economy transitions arguably support calls for Earth jurisprudence – where governance is Earth-centric rather than anthropogenic – including a rights of nature framework, which recognises law as a necessary tool for societal transformation (Kauffman and Martin, 2021; see also: Rühs and Jones, 2016; Graham and Maloney, 2019; Graham et al., 2024).

A feature of circular economy rhetoric has been the claim of unlocking multiple benefits for various stakeholders. Estimations of these include: realising a circular economy can result in 700,000 new jobs in the European Union (EU) by 2030 (Cambridge Econometrics et al., 2018), financial benefits are calculated to be around EUR 1.8 trillion by 2030 in Europe and USD 10 trillion by 2040 in China (Ellen MacArthur Foundation and McKinsey Centre for Business and Environment, 2015; Ellen MacArthur Foundation and Arup, 2018), and a circular economy can be critical in mitigating climate impacts, as 67 percent of global greenhouse gas emissions are related to material management (Circle Economy, 2020).

Despite the touted benefits, the circular economy concept is currently often implemented in a way that reinforces the status quo and business-as-usual. Action claiming to pursue circular economy transitions commonly focus on recycling (rather than reducing resource use and preventing waste) and the creation of markets for such recycled secondary materials (Steenmans and Lesniewska, 2023). If the concept is used in a sustained, tokenistic manner without impacting practices and behaviours, it risks becoming a marketing tool in greenwashing activities (i.e. the dissemination of misleading environmental claims; see de Freitas Netto et al., 2020; Konietzko et al., 2023; Steenmans and Lesniewska, 2023). Moreover, circular economy transitions are neither occurring at the speed nor scale necessary to cause systemic change (European Court of Auditors, 2023; Lesniewska and Steenmans, 2023). Current approaches are insufficient and additional interventions are needed to facilitate the necessary changes. Technology is recognised in circular economy laws and policies as one such possible broad enabling lever of change (see Section 2.2). Similarly, "circular economy advocates frequently identify digitalisation … as fundamental to rapidly upscaling circularity across sectors at all levels" (Lesniewska and Steenmans, 2023, p. 144). Combined, all this rhetoric boosts perceptions of blockchain's significant role as an enabling instrument towards fostering the circular economy (França et al., 2020; Narayan and Tidström, 2020; Böckel et al., 2021; Kuae and Machado, 2021; Upadhyay et al., 2021; Gong et al. 2022; Basile et al., 2023; see also Section 2).

To facilitate debate on blockchain's future beyond high-level enthusiasm and instead about responsible innovation choices, more detailed exploration of the pathways by which blockchain would contribute to the ultimate circular transformation of global resource systems is needed. As we write this in what is likely still an early stage of blockchain technology development, other scholars and practitioners should engage urgently in exploring the possible, diverse pathways towards a future of greater blockchain use. Only through such broader deliberation can our assumptions and mental model shaping possibly change, as well as our understanding of the associate benefits and risks be challenged to look beyond the status quo.

The aim of this chapter is therefore not to provide readers with predictive, or even prescriptive, outlines of pathways for blockchain-contributed change towards circular economies. Rather, we seek to present a set of illustrative outlines of blockchain-related pathways of change towards circular economies as exploratory prompts to provoke others in their thinking and imagination of alternative, sustainable futures. The next section therefore first introduces *what* uses have been identified to date for blockchain in enabling circular economy transitions. We draw on scholarly literature and observed experiences from blockchain applications in practice to summarise current uses of blockchain towards circular economy transitions, the challenges faced, as well as the specific role of laws and policies in supporting these uses. Section 3 then presents four pathways offering richer description of *how* blockchain will support laws and policy in effective transformation towards global circular economy systems. Finally, Section 4 briefly reflects on areas for further work to inform our understanding of blockchain's transformative potential in the context of the circular economy.

2. Blockchain and A Circular Economy

Blockchain has been recognised by scholars as an enabling technology in unlocking the circular economy's benefits (e.g. França et al., 2020; Narayan and Tidström, 2020; Taylor et al., 2020; Böckel et al., 2021; Kuae and Machado, 2021; Steenmans et al., 2021a; Upadhyay et al., 2021; Gong et al. 2022; Basile et al., 2023). The primary inherent blockchain characteristics typically recognised as foundational to that potential are decentralisation and transparency (Narayan and Tidström, 2020; Roschko and Peterson, 2022; Upadhyay et al., 2022; Basile et al., 2023). These enable blockchain to provide certain benefits for circular economies, such as the extraction of provenance of resources and wastes in supply chains, and enhanced trustworthiness of such information (França et al., 2020; Böckel et al., 2022; Cao et al., 2022; Powell et al., 2023). In turn, these can contribute to the circular economy through improved information sharing, potentially reducing transaction costs, enhancing performance, aiding resource efficiency, and reducing the carbon footprint along supply chains (Böckel et al., 2021; Upadhyay et al., 2021; Roschko and Peterson, 2022).

In contrast to several theorised benefits, three main functions of blockchain technology have primarily been the focus of practical efforts towards catalysing circular economy approaches and experiments (Steenmans et al., 2021a):

1. *Cryptocurrency payments (including cryptocurrency-based reuse and recycling rewards)* (Steenmans et al., 2021a; Gong et al., 2022): This is the use of cryptocurrency payments for transactions promoting circular practices of reuse, recycling, or other recovery. Application of this category is not limited to circular approaches, as such payments can equally be used in linear take-make-waste production and consumption processes. A sub-category specific to circularity is cryptocurrency-based reuse and recycling rewards, which in essence aim to directly incentivise behaviour of users by rewarding certain 'circular' behaviours, such as bringing in waste items for reuse or recycling (Steenmans et al., 2021a; Smajgl and Schweik, 2022).
2. *Monitoring and tracking of waste* (Kuae and Marcelo, 2021; Steenmans et al., 2021a; Walden et al., 2021; Gong et al., 2022): Blockchain can be used to capture data on resource and waste transactions, such as the provenance of resources and wastes, the amount of waste collected, the types of waste collected, etc. This information can then be used to target and support laws and policies aimed at minimising certain resource use and facilitating recycling through improved sorting.
3. *Smart contract implementation* (Steenmans et al., 2021a): Whereas the previous categories are application-based, smart contracts are primarily an implementation choice; the previous two categories can be implemented using smart contracts. In its essence, a smart contract is a computer program and data that can be used to digitally execute, monitor, or enforce agreements (Natanelov et al., 2022). The automation of such transactions can minimise administrative burdens and improve cost efficiency, which can promote circular approaches (Steenmans et al., 2021a).

2.1 Challenges

Multiple challenges exist in blockchain-based initiatives for the circular economy. Before implementation, there are potentially significant upfront costs involved in implementing the technology (Upadhyay et al., 2021). This perhaps in part explains why adoption of blockchain for circular economies is still mostly at the demonstration and piloting stages (Kouhizadeh et al., 2020; Kuae and Machado, 2021; Steenmans et al., 2021a). Evidence from the limited experiences to date suggest that the main challenges are related to infrastructure, including failures of interoperability, technological security, and stability issues, as well as issues with discrepancies between management ability and technology development (Kouhizadeh et al., 2020; Jiang et al., 2023). There are also concerns of 'garbage in, garbage out'; the challenge of ensuring only honest and correct data is put on the blockchain (Powell et al., 2022; see Sedlmeir et al., 2022 for an example of such challenges in relation to green energy labelling and certification). Undoubtedly further challenges will still be uncovered as blockchain-based applications to circular approaches are more widely adopted.

Blockchain has also been criticised for its intensive energy usage and resultant detrimental environmental impacts—often citing the computationally expensive proof-of-work consensus algorithm used in Bitcoin (Steenmans et al., 2021a). This energy footprint would neutralise any of the positive contributions to circular transitions, as sustainable energy consumption is an essential condition for achieving a circular economy (Steenmans, 2021). Conversely, there are increasingly more studies with evidence to support cases that blockchain can improve environmental sustainability (Ahl et al., 2022; Parmentola et al., 2022; Tawiah et al., 2022), particularly when alternative approaches to consensus are applied in projects with an environmental focus (Powell et al., 2021).

Care should be taken to avoid the framing of blockchain technology as a silver bullet for achieving circular economy objectives (Steenmans et al., 2021a, 2021b). Blockchain should not be heralded as a 'quick fix' for facilitating circular economy transitions, as it simplifies the complex challenges thereof and results in technological solutionism (Morozov, 2013). It is not even a necessary tool, as the benefits of blockchain can be achieved through other means, e.g. other data ledgers exist that can facilitate information sharing. Moreover, where adopted, blockchain needs to be complemented with other technologies, as it does not, for example, result in sustainable design or in maximising product use (Böckel et al., 2021; Steenmans et al., 2021b). Nonetheless, blockchain can make valuable contributions, as it can operationalise some characteristics that contribute to good governance and facilitate multi-actor engagement (Steenmans et al., 2021a; see Section 3 for additional contributions).

This section provides only a snapshot of some of the challenges that exist. There are other general issues of blockchain technology, such as security issues (hackers and show dealing) and public perception, which are not specific to circular resource and waste management applications (Jiang et al., 2023). Further discussion of general challenges is available elsewhere in this book in relation to other applications and contexts, as well as other scholarly discussions (e.g., Ziolkowski et al., 2020; Ye at al. 2021). Having set out the wider context of opportunities and challenges for blockchain technology facilitating circular economy transitions, we discuss the role of law and policy within this context in the following subsection.

2.2 The Role of Law and Policy

Amongst the myriad of individual and collective levers for governing the alignment of technological innovation and sustainable development, law and policy offer a critical set of powerful instruments to influence and steer the behaviours of diverse stakeholder groups and, in the long-term, strengthen or weaken some of their underlying mechanisms of change.

Within the legal landscape, blockchain is generally not yet explicitly linked to the circular economy (Steenmans et al., 2021b). Across the laws examined by Steenmans et al. (2021b), only the EU's non-legally binding Circular Economy Action Plan[2] explicitly refers to blockchain technology to lend support that blockchain can help "accelerate circularity," but without much specificity. Other laws only identify a role for technology generally (Steenmans et al., 2021b). The absence of a link is mirrored in practice and applications: across 21 initiatives examined by Steenmans et al. (2021a), only one initiative, which is now defunct, was developed with an explicit claim to support the implementation of laws.

Simultaneously, laws may inhibit blockchain adoption. Barriers identified include: legal uncertainty, data protection rules, ineffectiveness of current laws, and lack of regulation (Steenmans et al., 2021a; Cao et al., 2022; Jiang et al., 2023).

Yet, the law is necessary: the three current categories of blockchain-based applications discussed earlier can be implemented in a way that is contrary to circular approaches. Complementary targets and objectives are therefore needed in laws to guide the way in which blockchain is adopted. There may also be an increase in importance of the role of technology within laws as a result of increased complexity in supply chains and focus on technological development. For example, the proposed EU directive on common rules promoting the repair of goods requires information to be made available to consumers (European Commission, 2023). It does not clarify how, but a technology like blockchain is needed to record the significant amount of information that is required to support monitoring of this proposed directive. Blockchain can also be used to support compliance with legal tools such as extended producer responsibility, bans, and taxes, as well as to facilitate a plastic credits market similar to carbon credits (Kuae and Machado, 2021).

In the next section we bring together the previous high-level summaries of *what* uses have been imagined for blockchain in circular economy projects, to richer description of *how* blockchain will interlink with requisite legal and policy actions to facilitate circular transitions.

3. Blockchain's Transformative Potential for Circular Economy Transitions

Building on our previous work, which is based on empirical research including a review of more than 20 blockchain technology-based initiatives and interviews (Taylor et al. 2020; Steenmans et al., 2021a, 2021b), we discuss four pathways of change for illuminating blockchain's transformative potential for circular economy transitions: (1) facilitating equitable participation, (2) reframing rights, (3) distributing responsibility, and (4) transgressing boundaries of public and private law spheres.

---

[2] European Commission. (2020). *A New Circular Economy Action Plan*. COM(2020) 98 final. https://eur-lex.europa.eu/legal-content/EN/TXT/?qid=1583933814386&uri=COM:2020:98:FIN. See European Commission (2024) for an overview of the EU's circular economy plan.

3.1 Facilitating Equitable Participation

Blockchain technology-based initiatives facilitate multiple stakeholder interactions. In Steenmans et al. (2021a), we observed that public, private, and third sector actors implement such initiatives within waste management for the benefit of a multiplicity of actors, including government authorities, private sector organisations, as well as civil society. Blockchain can promote social inclusion of low-income and marginalised populations and the informal waste sector (França et al., 2020; Gong et al., 2022), in part through its open and distributed architecture, and its ability to monetise opportunities for informal economy workers to earn incomes from recycling (França et al., 2020; Gong et al., 2022). The Plastic Bank (plasticbank.com), for example, uses blockchain to securely execute plastic transactions that reward community members of vulnerable coastlines with bonuses that provide, inter alia, groceries, cooking fuel, school tuition, health insurance, digital connectivity, and fintech services in return for bringing in ocean-bound plastic waste (Plastic Bank, 2023a). This is supported by an app, which has a digital wallet that "is often the first savings account that [Plastic Bank's] members own" (Plastic Bank, 2023b).

Including the informal waste sector is critical, as they are a key contributor to recycling processes and are needed to deliver an inclusive, sustainable circular economy (Ezeah et al., 2013; Sengupta et al., 2022; Velis et al., 2022). The informal waste sector comprises stakeholders (individuals, groups, and micro-enterprises) not included in formal waste management operations, such as street waste-pickers who recover waste from the streets (Aparcana, 2017). The informal recycling sector alone, for example, is estimated to collect around 27.4 million tonnes of plastic (Lau et al., 2020; Velis et al., 2022). It, however, differs greatly between sectors and regions—the sector contributes significantly in both developing and transition countries (Ezeah et al., 2013), as well as other countries, such as in Norway where the informal sector in clothes reuse is "probably more important than the pecuniary markets" (Laitala and Klepp, 2017, p. 74). Furthermore, in many contexts and sectors, the formal and informal waste sectors are interdependent (Grant and Oteng-Ababio, 2016; Rendon et al., 2021). The waste management sector is thus a complex and sprawling landscape with a vast range of actors, which blockchain can support in mapping, understanding, and engaging with through facilitating monitoring and tracking.

Simultaneously, there are challenges to overcome to ensure blockchain can facilitate inclusion. de Jesus and Mendonça (2018) identify general technical barriers to circular economy transitions, including lack of sufficiently educated and specialised persons and other dimensions of digital inclusion such as access, affordability, and the ability to use digital platforms effectively. Implementing blockchain-based solutions may thus exacerbate such barriers by requiring a high level of technological literacy amongst stakeholders (Oberhauser, 2019). Otherwise, intermediaries are needed, which then negates the decentralised and empowering character of blockchain (McQueenie et al., 2024; Oberhauser, 2019).

In examining blockchain for environmental governance, particularly for payments for ecosystem services, Oberhauser (2019) concluded that:

> "it is unlikely that the decentralized and immutable character of blockchain smart contracts can be harnessed to address issues of equity and benefit distribution in complex constellations of communal natural resource management … This is mainly due to the inaccessibility of blockchain technology for the relevant stakeholders, which results in heavy reliance on trusted intermediaries. Thereby, the distinguishing properties of blockchain are largely nullified and genuinely participatory approach are ruled out" (pp. 13-14).

There are also justice concerns, and a delicate balance is required. Continuing to rely on initiatives, such as the Plastic Bank, deviates from environmental principles, such as the polluter-pays (i.e. the party responsible for producing the pollution is responsible for paying for managing the damage to human health or the environment) and proximity principles (i.e. waste should be managed as close as possible to its place of production), that are the basis for much environmental law. Yet, focusing on the prevention of waste may deprive the informal sector of their incomes. Some scholars therefore argue for the formalisation of the sector to help empower informal recyclers, improve efficiency, and living and working conditions (Wilson et al., 2006; Medina, 2008; Aparcana, 2017). Governance responses, however, need to be carefully crafted to ensure the sector is not criminalised and result in other issues (Goldstein, 2016). Further research is thus needed on how to include the informal sector in circular economy transitions and the possible role of a technology like blockchain—e.g. in mapping and involving the diverse actors needed to achieve sustainable and just circular resource and waste management—including considerations of resource, skills, and capacity needs. Lesniewska and Steenmans (2023) summarise this need as:

> "legal research on how digital technologies are incorporated into [circular economy] systems, and the equity and justice impacts such as access to material and wealth flows, benefit sharing, surveillance, and privacy, as well as digital exclusion and discrimination, need to be on the agenda" (p. 144).

3.2 Reframing Rights

Achieving a circular economy requires fundamental changes to the current linear economic system, and production and consumption processes. Some of the ways it is hoped such changes will be achieved are through challenging our current approaches to our use of products. We focus on two examples: (1) the right to repair to reduce product obsolescence and lengthen product life-cycles, and (2) rethinking current dominant approaches to property rights and ownership by shifting towards a sharing economy.

In a circular economy, objectives often include extending the lifespan of products and reducing planned obsolescence, i.e. where a product has been designed to become obsolete in order to drive extractivism, production, and consumption (Bedford et al., 2021). Right to repair laws and policies aim to tackle such planned obsolescence. They are increasingly adopted, with laws implemented in, for example (and not limited to), Australia, the EU, France, the UK, and Massachusetts and New York, USA. Within the EU, the European Commission has proposed a directive on common rules promoting the repair of goods (European Commission, 2023). This mandates that consumers are to be provided with a 'European Repair Information Form' (European Commission, 2023, Article 4), which requires information such as identity and contact details of the repairer and information on the repair service (European Commission, 2023, Annex I). A repairer is defined as:

> "any natural or legal person who, related to that person's trade, business, craft or profession, provides a repair service, including producers and sellers that provide repair services and repair service providers whether independent or affiliated with such producers or sellers" (European Commission, 2023, Article 2(2)).

There is thus a lot of information that needs to be accessible depending on the product. For example, if a refrigerator component breaks, is the repairer of the product the producer or seller of the broken component or seller of the refrigerator (if different), or a third party? The repairer would have to be clearly defined in, for example, the sales contract but it may depend on the particulars. Much information may therefore need to be stored on the different stakeholders related to various product components as well as the product in its entirety. Traceability offered by blockchain can support national implementation of the proposed directive by providing a database of the necessary provenance information required to facilitate repair. Hernandez et al. (2020) identify factors inhibiting repairability as: lack of knowledge on how products work, lack of spare parts, technical information, and restricted contracts, lack of economic incentives to repair a product, lack of emotional and economic attachment to products, and lack of design and manufacturing features allowing repairs. While identifying producers is not listed as an issue, such provenance could be a gateway to elucidating others, for example, providing mechanical details or listing useful contacts. Where repair is not possible, having information about contents also facilitates recyclability and recovery of product, as there can be many challenges in recycling unknown materials.

Additionally, a circular economy may entail more sharing. The sharing economy is broadly defined by Plewnia and Guenther (2017) as "activities or platforms which facilitate the sharing of material, products, product services, space, money, workforce, knowledge, or information based on for-profit or non-profit transactions in a variety of different market structures" (p. 576). The sharing and circular economy concepts are linked with the sharing economy considered a subset of the circular economy, though a better understanding of the relations between these two concepts is still needed (Henry et al., 2021). Libraries are an example of a sharing system, but demonstrate that distributed blockchain technology is neither relevant nor needed, as there is an obvious central authority (the library) and a history of libraries. There may be a use case for blockchain if there is a move away from a traditional library-approach sharing system, where the book is loaned and then returned, but instead an item may be passed along to 'loaners' (with the location accessible via the ledger).[3] A system is then needed to monitor the users. Furthermore, sharing systems can also require the pooling of information to facilitate sharing and systems for deposits. Again, blockchain is one possible implementation of such a system, to which there are many alternatives.

Discussions on both the right to repair and a sharing economy have alluded to a closely related issue discussed in the next section of distributing responsibility.

3.3 Distributing Responsibility

Whereas the benefits and limitations of the previous section focused on facilitating the enforcement of rights, this section is concerned with enforcing duties. In legal terms, these are closely related concepts: if X has a right (i.e. a 'claim' or 'privilege'), for example, to use Y's waste in a sharing economy, then the correlative and equivalent is that Y has a duty not to exclude X from using their waste (Hohfeld, 1913). With this link, it is unsurprising that many of the reasons why blockchain is a supportive tool for reframing rights also apply to why it can facilitate the distribution of responsibility.

Blockchain can facilitate tracking and monitoring, which is conducive to legal instruments providing for producers of products to retain some prescribed form of responsibility throughout their products' entire life-cycles. The right to repair discussed in the previous section, for example, can be construed as producers retaining responsibility for the repair of their products (if other legal conditions and requirements as set out in the relevant laws are met). Similarly, blockchain can facilitate extended producer responsibility (Taylor et al., 2020; Steenmans et al., 2021b; Gong et al., 2022). Extended producer responsibility can hold producers responsible throughout the life-cycle stages of their products, with the nature of responsibility being economic, physical, informative, or a form of liability. The specifics (i.e. type of responsibility and at what life-cycle stages) will depend on the particular law. To be able to implement, monitor, and enforce such responsibility effectively, the producer responsible has to be identifiable. Echoing the issues with right to repair, the complexity of some products' design means that blockchain may help support the recording of relevant actors throughout the product's life-cycle.

Many challenges, however, remain. There is, for example, uncertainty in some laws as to who has responsibility for repairing products, for taking back products, or for managing damage to human health or the environment as a result of products, etc. (e.g. Taylor et al., 2020; Bedford et al., 2021). Such issues need to be resolved

---

[3] The book may eventually be returned to the library, as in the example the library remains the owner of the book so can request its return, but it does not necessarily need to be returned.

before blockchain can be a supportive, facilitative, or even transformative tool for such laws.

3.4 Transgressing Boundaries of Public and Private Law Spheres

Current scholarly discussions on circular economy transitions predominantly focus on public law and policy tools, as well as law and policy approaches to circular economies are primarily situated within an area of primarily public law: environmental law (Lesniewska and Steenmans, 2023). Public law concerns the governance of relationships between public (e.g. national government and local authorities) and private entities (e.g. individual citizens and businesses). In contrast, "[p]rivate governance describes the various forms of private enforcement, self-governance, self-regulation, or informal mechanisms that private individuals, companies, or clubs (as opposed to government) use to create order, facilitate exchange, and protect property rights" (Stringham, 2015, pp. 3-4). Public law interventions are, however, "neither causing the systemic shift necessary nor delivering change at the pace needed to tackle current resource and waste crises" (Steenmans and Ulfbeck, 2023, p. 1).

Private law tools (e.g. within contract law, consumer protection, property rights) can provide additional and complementary opportunities for circular economy transitions (Mak and Terryn, 2019; Ballardini et al., 2021; Lesniewska and Steenmans, 2023; Steenmans and Ulfbeck, 2023).

Blockchain can facilitate private law interventions. Cowen (2020) sets out three particular contributions blockchain technology makes to supporting private governance: (1) blockchain offers a new way of tackling the power positions between the governor and the governed through providing a shared environment, (2) decentralised systems can be designed to be somewhat resistant to entanglement with existing political authorities, and (3) blockchain can facilitate rapid competition between rulesets, as they rely on very little infrastructure or human administrators for efficacy.

There are, however, limits to the extent to which blockchain can support private governance:

> "what blockchains offer is a new mechanism by which a lawgiver can selectively alienate themselves from some of their power and thus commit to predate off their subordinates in the future ... But ... as it stands, changing governing rules either implies violent revolution or costly contestation with a democratic process. Neither approaches look anything like a competitive market and offer little guarantee of improvement over a *status quo*. In essence, blockchains are a technology for lowering at some of the formidable transaction costs associated with moving from one set of rules to another" (Cowen, 2020, p. 215).

Blockchain thus helps transgress the public-private law boundaries by promoting private approaches, but does not remove the need for public law. Blockchain initiatives can be implemented in ways that do not support a sustainable, inclusive circular economy (Steenmans et al., 2021a). There is thus still a need to have complementary public interventions setting, for example, overarching objectives, targets, standards, etc.

Conclusion

Blockchain technology has the potential to contribute to multiple areas of legal and policy innovation supportive of attaining global and local circular economies. We have illustrated that the specifics of *how* blockchain might influence such change extends across a diverse set of transformation pathways encompassing different areas of economic and social innovations. Blockchain may, for example, facilitate new distributions of responsibility,[4] or connect diverse actors transgressing traditional, artificially-placed legal boundaries.

The value of exploring and reporting alternative transformation pathways is not determined by their eventual accuracy in predicting future blockchain uses. Instead, their value derives from the way they currently make explicit some of our working assumptions and understanding of blockchain's causal potential to contribute to transformative change. They provide a device within which we can explore the ways by which industry, society, and planet agendas might differentially interlink following contrasting blockchain usage choices. These pathways of change therefore provide a means, not an end. Our aim was not to present readers with a definitive or exhaustive list of pathways describing anticipated blockchain-enabled transformation, but rather to provoke readers into thinking about pathways likely familiar to them, and potentially speculate on the implications for the future should some or all unfold as described. Crucial next steps are deeper exploration still about the different risks, opportunities, harms, and benefits these or other pathways may generate for different stakeholder groups. And ultimately, collective reflection and deliberation is needed on which future blockchain-enabled transformation pathways we therefore want to make more or less possible.

References

Ahl, A., Goto, M., Yarime, M., Tanaka, K. and Sagawa, D. (2022). Challenges and opportunities of blockchain energy applications: Interrelatedness among technological, economic, social, environmental, and institutional dimensions. *Renewable and Sustainable Energy Reviews*, *166*, e112623. https://doi.org/10.1016/j.rser.2022.112623.

Aparcana, S. (2017). Approaches to formalization of the informal waste sector into municipal solid waste management systems in low- and middle-income countries: Review of barriers and success factors. *Waste Management*, *61*, 593–607. https://doi.org/10.1016/j.wasman.2016.12.028.

Ballardini, R.M., Kaisto, J. and Similä, J. (2021). Developing novel property concepts in private law to foster the circular economy. *Journal of Cleaner Production*, *279*, e123747. https://doi.org/10.1016/j.jclepro.2020.123747.

---

[4] For a discussion of what innovation in the context of sustainable development, see Ciarli (2022).

Basile, D., D'Adamo, I., Goretti, V. and Rosa, P. (2023). Digitizing circular economy through blockchains: The Blockchain Circular Economy Index. *Journal of Industrial and Production Engineering*, *40*(4), 233–45. https://doi.org/10.1080/21681015.2023.2173317.

Bedford, L., Mann, M., Foth, M. and Walters, R. (2022). A post-capitalocentric critique of digital technology and environmental harm: New directions at the intersection of digital and green criminology. *International Journal for Crime Justice and Social Democracy*, 11(1), 167–181. https://doi.org/10.5204/ijcjsd.2191

Böckel, A., Nuzum, A.-K. and Weissbrod, I. (2021). Blockchain for the circular economy: Analysis of the research-practice gap. *Sustainable Production and Consumption*, *25*, 525–39. https://doi.org/10.1016/j.spc.2020.12.006.

Cambridge Econometrics, Trinomics, & ICF (2018). *Impacts of circular economy policies on the labour market*. European Commission. https://circulareconomy.europa.eu/platform/sites/default/files/ec_2018_-_impacts_of_circular_economy_policies_on_the_labour_market.pdf.

Cao, S., Boyen, X., Deane, F., Miller, T. and Foth, M. (2022). Enabling cross-border trade in the face of regulatory barriers to data flow – the case of the blockchain-based service network. In A. V.S., A. S., J. Goldston, & S. Williams (Eds.), *Blockchain for Industry 4.0* (pp. 285–303). CRC Press. https://doi.org/10.1201/9781003282914-15.

Cao, S., Foth, M., Powell, W., Miller, T. and Li, M. (2022). A blockchain-based multisignature approach for supply chain governance: A use case from the Australian beef industry. *Blockchain: Research and Applications*, *3*(4), 100091. https://doi.org/10.1016/j.bcra.2022.100091.

Ciarli, T. (Ed.) (2022). *Changing directions: Steering science, technology and innovation towards the Sustainable Development Goas*. STRINGS, SPRU, University of Sussex. https://dx.doi.org/10.20919/FSOF1258.

Circle Economy (2020). *The Circularity Gap Report*. www.circle- economy.com/resources/circularity-gap-report-2020#:~:text=The%20Circularity%20Gap%20Report%202020,was%20first%20launche d%20in%202018.

Cowen, N. (2020). Markets for rules: The promise and peril of blockchain distributed governance. *Journal of Entrepreneurship and Public Policy, 9*(2), 213–26. https://doi.org/10.1108/JEPP-03-2019-0013.

de Freitas Netto, S.V., Sobral, M.F.F., Ribeiro, A.R.B. and Soares, G.R. da L. (2020). Concepts and forms of greenwashing: a systematic review. *Environmental Sciences Europe*, *32*(1), 1–12. https://doi.org/10.1186/s12302-020-0300-3.

de Jesus, A. and Mendonça (2018). Lost in transition? Drivers and barriers in the eco- innovation road to the circular economy. *Ecological Economics*, *145*, 75–89. https://doi.org/10.1016/j.ecolecon.2017.08.001.

Ellen MacArthur Foundation and McKinsey Center for Business and Environment (2015). *Growth within: A circular economy vision for a competitive Europe*. www.ellenmacarthurfoundation.org/growth-within-a-circular-economy-vision-for-a- competitive-europe.

Ellen MacArthur Foundation, & Arup (2018). *The circular economy opportunity for uban and industrial innovation in China*. www.ellenmacarthurfoundation.org/urban-and- industrial-innovation-in-china.

European Commission (2023). Proposal for a Directive of the European Parliament and of the Council on common rules promoting the repair of goods and amending Regulation (EU) 2017/2394, Directives (EU) 2019/771 and (EU) 2020/1828. COM(2023) 155 final. https://eur-lex.europa.eu/legal-content/EN/TXT/?uri=CELEX%3A52023PC0155.

European Commission. (2024). Circular https://environment.ec.europa.eu/strategy/circular-economy-action-plan_en.

European Court of Auditors (2023). *Circular economy: Slow transition by member states despite EU action* (Special Report 17). www.eca.europa.eu/en/publications/SR-2023-17.

Ezeah, C., Fazakerley, J.A. and Roberts, C.L. (2013). Emerging trends in informal sector recycling in developing and transition countries. *Waste Management*, *13*(11), 2509–19. https://doi.org/10.1016/j.wasman.2013.06.020.

França, A.S., Amato Neto, J., Gonçalves, R.F. and Almeida, C.M.V.B. (2020). Proposing the use of blockchain to improve the solid waste management in small municipalities. *Journal of Cleaner Production*, *244*, eArticle 118529. https://doi.org/10.1016/j.jclepro.2019.118529.

Friant, M.C., Vermeulen, W.J.V. and Salomone, R. (2020). A typology of circular economy discourses: Navigating the diverse visions of a contested paradigm. *Resources, Conservation and Recycling*, *161*, e104917. https://doi.org/10.1016/j.resconrec.2020.104917.

Global Footprint Network (2018). Earth Overshoot Date 2018 is August 1. https://www.footprintnetwork.org/2018/07/23/earth-overshoot-day-2018-is-august-1-the-earliest-date-since-ecological-overshoot-started-in-the-early-1970s-2/.

Goldstein, J. (2016). A pyrrhic victory? The limits to the successful crackdown on informal-sector plastics recycling in Wenan County, China. *Modern China*, *43*(1), 3-35. https://doi.org/10.1177/0097700416645882.

Gong, Y., Xie, S., Arunachalam, D., Duan, J. and Luo, J. (2022). Blockchain-based recycling and its impact on recycling performance: A network theory perspective. *Business Strategy and the Environment*, *31*(8), 3717–41. https://doi.org/10.1002/bse.3028.

Graham, M. and Maloney, M. (2019). Caring for country and rights of nature in Australia: A conversation between earth jurisprudence and aboriginal law and ethics. In C. La Follette & C. Maser (Eds.), *Sustainability and the Rights of Nature in Practice* (pp. 385–99). CRC Press. https://doi.org/10.1201/9780429505959-19.

Graham, M., Maloney, M. and Foth, M. (2024). A City of Good Ancestors: Urban Governance and Design from a Relationist Ethos. In S. Heitlinger, M. Foth, & R. Clarke (Eds.), *Designing More-than-Human Smart Cities: Beyond Sustainability, Towards Cohabitation* (pp. 239–66). Oxford University Press. https://doi.org/10.1093/9780191980060.003.0014.

Grant, R.J. and Oteng-Ababio, M. (2016). The global transformation of materials and the emergence of informal urban mining in Accra, Ghana. *Africa Today*, *62*(4), 3–20. https://doi.org/10.2979/africatoday.62.4.01.

Henry, M., Schraven, D., Bocken, N., Frenken, K., Hekkert, M. and Kirchherr, J. (2021). The battle of the buzzwords: A comparative review of the circular economy and the sharing concepts. *Environmental Innovations and Societal Transitions*, *38*, 1–21. https://doi.org/10.1016/j.eist.2020.10.008.

Hohfeld, W. N. (1913). Some fundamental legal conceptions as applied in judicial reasoning. *The Yale Law Journal*, *23*(1), 16–59. https://doi.org/10.2307/785533.

Jäger-Roschko, M. and Petersen, M. (2022). Advancing the circular economy through information sharing: A systematic literature review. *Journal of Cleaner Production*, *369*, e133210. https://doi.org/10.1016/j.jclepro.2022.133210.

Jiang, P., Zhang, L., You, S., Fan, Y.V., Tan, R.R., Klemes, J.J. and You, F. (2023). Blockchain technology applications in waste management: Overview, challenges and opportunities. *Journal of Cleaner Production*, *421*, e138466. https://doi.org/10.1016/j.jclepro.2023.138466.

Kauffmann, C.M. and Martin, P.L. (2021) *The politics of rights of nature. Strategies for building a more sustainable future*. MIT Press Direct. https://doi.org/10.7551/mitpress/13855.001.0001.

Kouhizadeh, M., Zhu, Q. and Sarkis, J. (2020). Blockchain and the circular economy: Potential tensions and critical reflections from practice. *Production, Planning & Control*, *31*(11–12), 950–66. https://doi.org/10.1080/09537287.2019.1695925.

Konietzko, J., Das, A. and Bocken, N. (2023). Towards regenerative business models: A necessary shift? *Sustainable Production and Consumption*, *38*, 372–88. https://doi.org/10.1016/j.spc.2023.04.014.

Kuae, H.O. and Machado, M. (2021). How can reciChain consortium project foster circular economy with blockchain? The ISPIM Innovation Conference, Germany.

Laitala, K., & Klepp, I. G. (2017). Clothing reuse: The potential in informal exchange. *Clothing Cultures, 4*(1), 61–77. https://doi.org/10.1386/cc.4.1.61_1.

Lau, W.W.Y., Shiran, Y., Bailey, R.M., Cook, E., Stuchtey, M.R., Koskella, J., Velis, C.A., Godfrey, L., Boucher, J., Murphy, M.B., Thompson, R.C., Jankowska, E., Castillo, A.C., Pilditch, T.D., Dixon, B., Koerselman, L., Kosior, E., Favoino, E., Gutberlet, J., Baulch, S., Atreya, M.E., Fischer, D., He, K.K., Petit, M.M., Sumaila, U.R.., Neil, E., Bernhofen, M.V., Lawrence, K. and Palardy, J.E. (2020). Evaluating scenarios towards zero plastic pollution. *Science*, *369*(6510), 1455–1461. https://doi.org/10.1126/science.aba9475.

Lesniewska, F. and Steenmans, K. (2023). *Circular economy and the law: Bringing justice into the frame*. Routledge. https://doi.org/10.4324/9780429355141.

Mak, V. and Terryn, E. (2019). Circular economy and consumer protection: The consumer as a citizen and the limits of empowerment through consumer law. *Journal of Consumer Policy*, *43*, 227–248. https://doi.org/10.1007/s10603-019-09435-y.

McCauley, D. and Heffron, R. (2018). Just transition: Integrating climate, energy and environmental justice. *Energy Policy*, *119*, 1–7. https://doi.org/10.1016/j.enpol.2018.04.014.

McQueenie, J., Foth, M. and Hearn, G. (2024). *Community, Culture, Commerce: The Intermediary in Design and Creative Industries*. Palgrave Macmillan. https://doi.org/10.1007/978-981-99-7889-2.

Medina, M. (2008). *The informal recycling sector in developing countries* (Gridlines No. 44). The World Bank. https://openknowledge.worldbank.org/server/api/core/bitstreams/30890fc0-a6ee-5cb0-8e91-c1157bb2a2d1/content.

Morozov, E. (2013). *To save everything, click here.* Penguin Books.

Narayan, R. and Tidström, A. (2020). Tokening coopetition in a blockchain for a transition to circular economy. *Journal of Cleaner Production*, *263*, e121437. https://doi.org/10.1016/j.jclepro.2020.121437.

Natanelov, V., Cao, S., Foth, M. and Dulleck, U. (2022). Blockchain smart contracts for supply chain finance: Mapping the innovation potential in Australia-China beef supply chains. *Journal of Industrial Information Integration*, 30(100389), 100389. https://doi.org/10.1016/j.jii.2022.100389

Oberhauser, D. (2019). Blockchain for environmental governance: Can smart contracts reinforce payments for ecosystem services in Namibia? *Frontiers in Blockchain*, 2. https://doi.org/10.3389/fbloc.2019.00021.

Parmentola, A., Petrillo, A., Tutore, I. and De Felice, F. (2022). Is blockchain able to enhance environmental sustainability? A systematic review and research agenda form the perspective of Sustainable Development Goals (SDGs). *Business Strategy and the Environment*, *31*(1), 194–17. https://doi.org/10.1002/bse.2882.

Plastic Bank (2023a). *If you could stop plastic form entering the ocean, would you?* https://plasticbank.com/about.

Plastic Bank. (2023b). *About Plastic Bank.* https://plasticbank.com/about-plastic-bank/Plewnia, F. and Guenther, E. (2017). Mapping the sharing economy for sustainability research. *Management Decision*, *56*(3), 570–83. https://doi.org/10.1108/MD-11-2016-0766.

Powell, W., Cao, S., Miller, T., Foth, M., Boyen, X., Earsman, B., del Valle, S. and Turner-Morris, C. (2021). From premise to practice of social consensus: How to agree on common knowledge in blockchain-enabled supply chains. *Computer Networks*, *200*, 108536. https://doi.org/10.1016/j.comnet.2021.108536.

Powell, W., Cao, S., Foth, M., He, S., Turner-Morris, C. and Li, M. (2023). Revisiting Trust in Supply Chains: How Does Blockchain Redefine Trust? In A. Bouras, I. Khalil, & B. Aouni (Eds.),

Blockchain Driven Supply Chains and Enterprise Information Systems (pp. 21–42). Springer International Publishing. https://doi.org/10.1007/978-3-030-96154-1_2.

Powell, W., Foth, M., Cao, S. and Natanelov, V. (2022). Garbage in garbage out: The precarious link between IoT and blockchain in food supply chains. *Journal of Industrial Information Integration, 25*, 100261. https://doi.org/10.1016/j.jii.2021.100261.

Rendon, M., Espluga-Trenc, J. and Verd, J.-M. (2021). Assessing the functional relationship between the formal and informal waste systems: A case-study in Catalonia (Spain). *Waste Management, 131*, 483–90. https://doi.org/10.1016/j.wasman.2021.07.006.

Richardson, K. and Bai, X. (2023, September 19). *What are 'planetary boundaries' and why should we care?* The Conversation. https://theconversation.com/what-are-planetary- boundaries-and-why-should-we-care-213762.

Richardson, K., Steffen, W., Lucht, W., Bendtsen, J., Cornell, S.E., Donges, J.F., Drüke, M., Fetzer, I., Bala, G., Von Bloh, W., Feulner, G., Fiedler, S., Gerten, D., Gleeson, T., Hofmann, M., Huiskamp, W., Kummu, M., Mohan, C., Nogués-Bravo, D., Petri, S., Porka, M., Rahmstorf, S., Schaphoff, S., Thonicke, K., Tobian, A., Virkki, V., Wang-Erlandsson, L., Weber, L. and Rockström, J. (2023). Earth beyond six of nine planetary boundaries. *Science Advances, 9*(37). https://doi.org/10.1126/sciadv.adh2458.

Rockström, J., Steffen, W., Noone, K., Persson, Å., Chapin, F.S. III, Lambin, E., Lenton, T.M., Scheffer, M., Folke, C., Schellnhuber, H.J., Nykvist, B., de Wit, C.A., Highes, T., van der Leeuw, S., Rodhe, H., S¨rlin, S., Snyder, P.K., Costanza, R., Svedin, U., Falkenmark, M., Karlberg, L., Corell, R.W., Fabry, V.J., Hansen, J., Walker, B., Liverman, D., Richardson, K., Crutzen, P. and Foley, J. (2009). A safe operating space for humanity. *Nature, 461*, 472–75. https://doi.org/10.1038/461472a.

Rühs, N. and Jones, A. (2016). The Implementation of Earth Jurisprudence through Substantive Constitutional Rights of Nature. *Sustainability: Science Practice and Policy, 8*(2), 174. https://doi.org/10.3390/su8020174.

Schröder, P. (2020). *Promoting a just transition to an inclusive circular economy* (Research Paper). Chatham House. www.chathamhouse.org/2020/04/promoting-just-transition-inclusive-circular-economy.

Sedlmeir, J., Völter, F. and Strüker, J. (2022). The next stage of green electricity labeling: using zero-knowledge proofs for blockchain-based certificates of origin and use. *SIGENERGY Energy Inform. Rev.* 1, 1, 20–31. https://doi.org/10.1145/3508467.3508470.

Sengupta, D., Ilankoon, I.M.S.K., Kang, K.D. and Meng, N.C. (2022). Circular economy and household e-waste management in India: Integration of formal and informal sectors. *Minerals Engineering, 184*, e107661. https://doi.org/10.1016/j.mineng.2022.107661.

Smajgl, A. and Schweik, C. M. (2022). Advancing sustainability with blockchain-based incentives and institutions. *Frontiers in Blockchain*, 5. https://doi.org/10.3389/fbloc.2022.963766.

Steenmans, K. (2021). The role of circular economy transitions in fostering sustainable energy democracy. In R. Fleming, K. Huhta, & L. Reins (Eds.), *Sustainable energy democracy and the law* (pp. 144–68). Brill Nijhoff. https://doi.org/10.1163/9789004465442_008.

Steenmans, K. and Lesniewska, F. (2023). Limitations of the circular economy concept in law and policy. *Frontiers in Sustainability*, 4. https://doi.org/10.3389/frsus.2023.1154059.

Steenmans, K. and Ulfbeck, V. (2023). Fostering the circular economy through private law: Perspectives from the extended producer responsibility concept. *Resources, Conservation and Recycling, 195*, e107016. https://doi.org/10.1016/j.resconrec.2023.107016.

Steenmans, K., Taylor, P. and Steenmans, I. (2021a). Blockchain technology for governance of plastics: Where are we? *Social Sciences, 10*(11), Article 434. https://doi.org/10.3390/socsci10110434.

Steenmans, K., Taylor, P. and Steenmans, I. (2021b) Regulatory opportunities and challenges for blockchain adoption for circular economies. *Proceedings of 2021 IEEE International*

Conference on Blockchain (Blockchain), Australia, 572–77. https://doi.org/10.1109/Blockchain53845.2021.00086.

Stringham, E.P. (2015). *Private governance: Creating order in economic and social life*. Oxford University Press. https://doi.org/10.1093/acprof:oso/9780199365166.003.0001.

Tawiah, V., Zakari, A., Li, G. and Kyiu, A. (2022). Blockchain technology and environmental efficiency: Evidence from US-listed firms. *Business Strategy and the Environment*, *31*(8), 3757–68. https://doi.org/10.1002/bse.3030.

Taylor, P., Steenmans, K. and Steenmans, I. (2020). Blockchain technology for sustainable waste management. *Frontiers in Political Science*, 2. https://doi.org/10.3389/fpos.2020.590923.

Upadhyay, A., Mukhuty, S., Kumar, V. and Kazancoglu, Y. (2021). Blockchain technology and the circular economy: Implications for sustainability and social responsibility. *Journal of Cleaner Production*, *293*, e126130. https://doi.org/10.1016/j.jclepro.2021.126130.

Velis, C.A., Hardesty, B.D., Cottom, J.W. and Wilcox, C. (2022). Enabling the informal recycling sector to prevent plastic pollution and deliver an inclusive circular economy. *Environmental Science & Policy*, *138*, 20–25. https://doi.org/10.1016/j.envsci.2022.09.008.

Walden, J., Steinbrecher, A. and Marinkovic, M. (2021). Digital product passports as enabler of the circular economy. *Chemie Ingenieur Technik*, *93*(11), 1717–27. https://doi.org/10.1002/cite.202100121.

Wang, X. and Lo, K. (2021). Just transition: A conceptual review. *Energy Research & Social Science*, *82*, e102291. https://doi.org/10.1016/j.erss.2021.102291.

Wilson, D.C., Velis, C. and Cheeseman, C. (2006). Role of informal sector recycling in waste management in developing countries. *Habitat International*, *30*(4), 797–808. https://doi.org/10.1016/j.habitatint.2005.09.005.

Ye, C., Cao, W. and Chen, S. (2021). Security challenges of blockchain in Internet of Things: Systematic literature review. *Transactions on Emerging Telecommunications Technologies*, *32*(8), e4177. https://doi.org/10.1002/ett.4177.

Ziolkowski, R., Miscione, G. and Schwabe, G. (2020). Decision problems in blockchain governance: Old wine in new bottles or walking in someone else's shoes? *Journal of Management Information Systems*, *37*(2), 316–48. https://doi.org/10.1080/07421222.2020.1759974.

12 Blockchain Applications for Carbon Trading – An Overview of Current Developments in China

Yijia Hao[1*], Shoufeng Cao[2*] and Yong Xia[3]

1. Introduction

Carbon trading is a market-based approach involving the trading of credits (of some form) aimed at reducing greenhouse gases (GHGs) that contribute to climate change (Lohmann, 2010). An increasing awareness of the impact of GHGs, particularly carbon dioxide, on climate change has accelerated the global commitment to reducing GHG emissions (Yoro and Daramola, 2020). In 2000, the European Union (EU) put forward the first European climate change plan, vigorously developing and utilising new energy sources. It was followed by the establishment of the first carbon market in the EU with a global reach, in 2005 (de Larragán, 2010). Since then, several countries in Europe and the United States have developed their own carbon markets. The trading of carbon credits from reducing emissions offers a cost-effective solution for mitigating emissions and therefore has the potential to help nations and the world meet the goals for GHG emission reduction. As nations continue to focus on sustainable economic development and drive towards a net zero economy in response to the urgency of global climate changes, carbon markets are expected to grow rapidly in the years ahead. This is underpinned by a recent Mckinsey report, which found that global demand for voluntary carbon credits could increase by a factor of 15 by 2030 and a factor of 100 by 2050 (Blaufelder et al., 2021).

China, one of the world's largest carbon emitters, initiated its effort to explore emissions trading as a primary means for mitigating carbon emissions in 2011 (Yang et al., 2023). Compared with the European Union Emissions Trading System (EU ETS) and California carbon trading, China's carbon neutral plan and carbon trading construction started relatively late. Based on the experience of other

[1] College of Economics and Management, Xinjiang Agricultural University, Urumqi, Xinjiang 830052, China.

[2] School of Agriculture and Food Sustainability, The University of Queensland, Brisbane, QLD 4067, Australia.

[3] College of Economics and Management, Xinjiang Agricultural University, Urumqi, Xinjiang 830052, China.
Email: sean0991@qq.com

* Corresponding author: 1147328085@qq.com (JH); s.cao@uq.edu.au (SC) (first and second authors contributed equally)

countries, China has established national cross-regional carbon rights exchanges and launched supporting trading mechanisms and trading systems. China has been conducting several pilot carbon emissions trading programs in the first seven provinces and cities, including Beijing, Shanghai, Tianjin, Chongqing, Hubei, Guangdong, and Shenzhen (Wen et al., 2021). During the pilot phase from 2013 to 2020, the total volume of carbon transactions in China reached an accumulated sum of 237.18 million tons, demonstrating continuous growth. China's pilot carbon market is expected to expand in parallel with the "dual carbon" goals – achieving carbon peaking before 2030 and reaching carbon neutrality before 2060 – declared by President Xi Jinping at the United Nations General Assembly in September 2020. Despite this potential transition, China's pilot program faces several critical problems, including insufficient transparency of the trading system, low trading efficiency, imperfect trading data storage mechanisms, and a lack of legal constraints (Cao, 2020), which are significant barriers to the implementation of the carbon emissions trading system and impair the overall functionality of the system (Jiang et al., 2022).

To address these prominent challenges pertaining to carbon trading, China has endeavoured to embrace technological innovation for carbon trading. China's pursuit is not confined to applying traditional technologies but extends to the cutting-edge frontier of blockchain technology. This is exemplified by President Xi Jinping's resolute support for blockchain adoption, recognising it as a key driver for independent innovation of core technologies. On 27 May 2021, the Ministry of Industry and Information Technology (MIIT) and the Central Internet Information Office (CNIO) issued the *Guiding Opinions on Accelerating the Promotion of Blockchain Technology Application and Industrial Development* in various fields (Crawley, 2021). China has applied blockchain technology in multiple industry scenarios such as government services, supply chain finance, and so on (Chen et al., 2021), with carbon trading emerging as a promising scenario. Blockchain technology has several inherent characteristics that enhance transparency, traceability, and immutability (Powell et al., 2021) and therefore provides a stronger boost to empower carbon trading development. Government support, coupled with the recognition of blockchain's technical capability in supporting carbon neutrality (Qin et al., 2023), fuel China's enthusiastic embrace of blockchain technology for the national carbon trading market launched in January 2021 (Tan, 2022).

The application of blockchain-based carbon emissions trading represents an important exploration within China's digital economy and green and low-emission development pathway to a sustainable future in the context of "double carbon" goals. It also demonstrates China's technological dedication to combating climate change and creating a sustainable and ecologically responsible future not just for itself but for the world. This chapter aims to present an overview of China's carbon trading, and current progress and possible challenges regarding the integration of blockchain technology. It contributes to a better understanding of China's blockchain initiatives for carbon trading from a practical perspective.

Section 2 overviews the trends and challenges with carbon trading in China. It is followed by the role of blockchain technology for carbon trading in Section 3. After presenting China's current exploration of blockchain technology for carbon trading in Section 4, we reflect on challenges and considerations in Section 5. Section 6 concludes this chapter with synthesised results and implications and outlook limitations, prospects and future directions.

2. Carbon Trading in China: Trends and Challenges

This section outlines the evolution of carbon trading in China (Section 2.1) and potential challenges inhibiting its development in the years ahead (Section 2.2).

2.1 Carbon Trading Trends and Developments in China

China's carbon emissions trading has evolved over three stages. From 2005 to 2010, China's carbon trading featured an engagement in International Clean Development Mechanism (CDM) projects (Thomas et al., 2011). Through involvement in CDM projects, China acquired capital and technology from developed countries, like the EU, to reduce carbon emissions. In exchange, these developed countries gain greenhouse gas (GHG) "emission rights" from their Chinese counterparts to offset their own carbon emissions (Thomas et al., 2011). Before the European Union Carbon Trading Mechanism (EU-ETS) no longer accepted newly registered CDM projects from China in 2013, China had established ETS pilot programs in 2011 and approved Beijing, Shanghai, Tianjin, Chongqing, Hubei, Guangdong, and Shenzhen to practise the "cap and trade" ETS (Wen et al., 2021; Jiang et al., 2022). Seven emissions trading scheme (ETS) pilot projects were conducted between 2013 and 2014, which marks the initiation of the regional carbon emissions trading pilot stage. In 2016, Fujian province joined the ETS pilot. Table 12.1 details carbon emissions trading pilots across China's eight provinces and cities. The industries covered by the eight pilot regions include power, chemicals, building materials and other eight major energy-consuming industries. In terms of market scale, Hubei province and Guangdong province have larger values than those of other regions, with Hubei province leading in both the total volume and value of carbon transactions. The total volume of carbon transactions in China from 2013 to 2020 reached an accumulated sum of 237.18 million tons, demonstrating continuous growth. The cumulative turnover amounted to approximately 5.747 billion yuan (~896.09 million USD), with 2020 marking the peak of this phase at about 1.267 billion yuan (~197.66 million USD) (Qianzhan Industry Research Institute, 2021).

Following the pilot project implementation, the official release of the *National Carbon Emission Trading Management Measures (for Trial Implementation)* on February 1, 2021, signified the formal transition of China's carbon emission trading pilot into the national implementation stage. On 16 July 2021, China's ETS trading system was officially launched for online trading, where a "dual city" model was adopted with Shanghai Environment and Energy Exchange Ltd. responsible for the

Table 12.1 China's carbon emissions trading pilots in eight provinces and cities

Region	*Initiation*	*Main trading products*	*Threshold*
Shenzhen	18/6/2013	SZEA, CCER	3,000 tons of coal for industry; 10,000 m^2 for government offices and large public buildings
Shanghai	26/11/2013	SHEA, CCER	20,000 tons of coal for industry; 10,000 tons for non-industry
Beijing	28/11/2013	BJEA, BCER, CCER	5,000 tons of coal for emission control units
Guangdong	19/12/2013	GDEA, PHCER, CCER	20,000 tons of coal
Tianjin	26/12/2013	TJEA, CCER	more than 20,000 tons of coal
Hubei	12/4/2014	HBEA, CCER	60,000 tons of coal
Chongqing	9/6/2024	CQEA, CCER	20,000 tons of coal
Fujian	12/12/2016	FJEA, FFCER, CCER	50,000 tons of coal

Data source: Based on "China Carbon Finance Market Research 2021", "China Carbon Finance Market Research and Analysis Report" and other related materials.
Note: CCER: China Certified Emission Reduction; SZEA: Shenzhen Emissions Reduction; SHEA: Shanghai Emission Allowance; BJEA: Beijing Emission Allowance; BCER: Beijing Forest Carbon Emissions Reduction; GDEA: Guangdong Emission Allowance; PHCER: Pu Hui Certified Emission Reductions; TJEA: Tianjin Emission Allowance; HBEA: Hubei Emission Allowance; CQCER: Chongqing Certified Emissions Reduction; FJEA: Fujian Emission Allowance; FFCER: Fujian Forestry Certified Emission Reduction.

construction of the trading system and Carbon Emission Rights Registration and Clearing (Wuhan) Co., Ltd. (referred to as China Carbon Registration) responsible for the construction of the registration and settlement system. The first performance cycle of the national carbon emissions trading market successfully concluded by 31 December 2021. It is reported that the cumulative transaction volume of carbon emission quotas in China's national carbon market recorded 179 million tons, with a cumulative transaction volume of 7.661 billion yuan (~1.197 billion USD).[1] China's ETS has established itself as the world's largest carbon market, surpassing the European Union's by threefold in size (Busch, 2022). According to the Shanghai Environment and Energy Exchange, the national carbon emission trading market had an annual trading volume of 212 million tons of carbon emission quotas in 2023, with an annual transaction value of 14.444 billion yuan (~2.274 billion USD), and a daily average trading volume of 875,800 tons.[2] Considering the limited coverage of industries in China's current carbon market, both the trading volume and transaction amount are anticipated to experience exponential growth as more heavy industries and manufacturing sectors are incorporated into the trading system.

1 https://www.mee.gov.cn/ywgz/ydqhbh/wsqtkz/202212/P020221230799532329594.pdf
2 http://www.tanpaifang.com/tanjiaoyi/2023/1230/103516.html

2.2 Problems and Challenges

Despite the rapid development in trading volume and transaction amount, together with the vast potential for expansion, China's carbon trading faces several problems and challenges that can inhibit further expansion if not properly resolved. This subsection highlights three prominent challenges that require particular attention, which are (i) insufficient transparency and carbon information disclosure, (ii) absence of a trustworthy verification and regulation mechanism, and (iii) risk of double counting.

(i) Insufficient transparency and carbon information disclosure

While China's carbon trading market has seen a growing trend in enhancing the transparency of carbon information and corporate carbon information disclosure, the advancement of carbon information disclosure remains constrained due to the current development status and corporate awareness and value. When China set the "double carbon" goals during the *14th Five-Year Plan* period, it was explicitly proposed to promote carbon emission transparency and establish a unified information disclosure system. It has set specific requirements on relevant enterprises across the country, but these requirements are on a voluntary basis without enforcement. However, Chinese companies are not very enthusiastic about voluntarily disclosing carbon emission information. In 2020, only 9.3% of manufacturing companies with high carbon emissions among companies listed on the main board of the Shenzhen Stock Exchange disclosed carbon emission information (Zeng et al., 2021). Additionally, there is a limited awareness among most Chinese companies regarding carbon emission information disclosure (Zeng et al., 2021), thereby hindering improvements in carbon disclosure and trading transparency. The lack of disclosure and transparency in carbon information can lead to scepticism among market participants regarding the credibility of transactions. This, in turn, may adversely impact the carbon trading market's activities and undermine market confidence, potentially reducing market liquidity (Cui et al., 2016). Therefore, the traditional carbon trading market needs to solve the problem of information asymmetry between transaction parties.

(ii) Absence of a trustworthy verification and regulation mechanism

Accurate measurement and verification of carbon emission reductions can be complex and costly, which causes verification and regulation challenges. Verification in the carbon market is essential for its functionable operation, and accurate data is crucial to ensure effective and standardised operations (Zhang, 2015). Currently, the development of carbon verification and regulation in China is still in its early stages, facing issues such as the lack of unified carbon verification standards and incomplete carbon regulation guidelines. The qualifications of carbon verification agencies in the Chinese market vary across regions, which may affect the accuracy of verifying enterprise carbon emission data, leading to abnormal or inaccurate data (Wei and Yang, 2017). Enterprise carbon emission data requires third-party verification, as there may be a possibility of fraudulent activities or collusion for improper benefits. Additionally, the regulation of the carbon emission market faces such challenges as high costs and complex regulatory procedures. The current system lacks clear provisions for regulating the carbon market, making it challenging to effectively prevent trading risks (Chen, 2017).

(iii) Risk of double counting

When the same emission reduction is claimed and sold multiple times, this gives rise to the potential for double counting, which undermines the integrity of the carbon trading system (Ma and Duan, 2024). The risk of double counting becomes more likely when coordination among various environmental rights trading mechanisms is inadequate. China's emerging carbon trading market is particularly susceptible to this risk due to its early-stage development and the absence of a comprehensive regulatory framework for carbon accounting and reduction. The incomplete regulatory system enhances the likelihood of double counting, even in the presence of information disclosure. This overestimation compromises the accuracy of reported emission reductions, undermining the credibility and effectiveness of carbon markets. Repeated calculation, application, and sale of a significant volume of emission reductions can also impact the pricing of carbon credits, impeding the market's ability to reflect the genuine scarcity of emission reductions. Consequently, this leads to reduced trading efficiency, resource inefficiency, and sluggish development of the carbon market.

3. The Role of Blockchain Technology for Carbon Trading

Blockchain technology, originally associated with cryptocurrencies like Bitcoin, has now evolved into a multifaceted tool with transformative potential across various industries (Cao et al., 2023). Blockchain applications offer innovative ideas and solutions to enhance the efficiency of the existing carbon trading market. Within China's carbon trading context, scholars and practitioners have investigated the suitability of blockchain applications in various aspects of the carbon market, including collaborative governance, quota allocation, trading, and regulation. Zheng et al. (2022) proposed collaborative governance of carbon trading as an example to highlight the value of implementing blockchain technology in public sharing platforms. Yan and Li (2020) proposed the establishment of a blockchain-enabled carbon emissions trading platform to build a complete closed-loop process, from quota issuance to market trading, and comprehensive supervision. Others have delved into the technical framework or architecture of blockchain-based carbon trading models. Zhou et al. (2020) outlined a pathway to construct a more comprehensive blockchain carbon trading model. Ye et al. (2022) designed a blockchain system architecture for the carbon market, incorporating individuals into the scope of carbon trading and constructing a carbon market system of "public chain + side chain". Their system enables enterprise-level carbon trading on the public chain and individual-level carbon trading on the side chain. Tang et al. (2023) investigated blockchain applications from a regulatory perspective and proposed integrating blockchain technology to create a visualisation model, allowing regulators to easily check and verify transaction data in real-time, ensuring the authenticity and accuracy of transactions. While these studies lack practical insights, they have shown in principle the suitability of blockchain applications for carbon trading from either conceptual or technical perspectives.

Blockchain technology has the unique characteristics of high transparency, traceability, security, reliability, and decentralisation. Blockchain-enabled carbon trading has the potential to address various challenges prevalent in traditional carbon trading markets.

(i) Improved transparency

Blockchain's immutable ledger provides improved transparency by recording all transactions throughout carbon trading processes. This can ensure regulators and market participants maintain the integrity of the market. Leveraging blockchain technology, the carbon trading system can be built as an open and transparent ecosystem. In addition to cryptographic protection for the private and sensitive information of trading parties on a blockchain network (Zhu et al., 2017), the remaining data is accessible to all, constituting a decentralised shared ledger with recorded transaction information. This system enables a high level of transparency by ensuring carbon transaction information is open, visible, and accessible. Regulators and market participants can write, verify, and query relevant carbon transaction data, facilitating collective oversight to ensure compliance with operational behaviours. The improved transparency can ensure credible sharing of carbon trading data and, consequently, bolster the efficiency of carbon trading operations.

(ii) Verifiable traceability

Blockchain ensures the traceability of carbon credits throughout their entire lifecycle, preventing double-counting and ensuring they are utilised for their intended purpose. This traceability becomes especially crucial in cases of trading disputes between parties involved in a carbon emission rights transaction or during government supervision and inspection of the carbon trading process. A blockchain network comprises interconnected nodes, each storing identical business information within the same category, based on various carbon trading business requirements. The blockchain's structure, with a block ID pointer data at its head, establishes connections between nodes to form a "data chain." This enables sequential querying of relevant nodes, allowing for the retrieval of specific content related to carbon trading. The pointer information facilitates the identification of superior nodes, ensuring that multiple nodes create a cohesive data chain. This process enables precise tracking and standardisation of the entire life cycle of carbon trading, offering a comprehensive and verifiable traceability management system (Yan and Li, 2020). However, it cannot avoid the risk of garbage in and garbage out as blockchain ensures data immutability on-chain only (Powell et al., 2021). Therefore, it is necessary to improve the accuracy of carbon trading data at the registry point.

(iii) Enhanced security

Blockchain's cryptographic features enhance security of carbon transactions, reducing the risk of fraud and unauthorised access (Powell et al., 2021). Employing an asymmetric encryption algorithm, blockchain technology can ensure end-to-end security by safeguarding carbon trading information throughout the entire lifecycle of the trading process. Once carbon trading data is verified and added to the blockchain, it becomes immutable and permanently stored. It becomes impossible to tamper with or manipulate this data without required permission from other carbon trading users. Even in the event of malicious tampering, the public keys released by both carbon trading users allow for scrutiny. This characteristic ensures that blockchain-enabled carbon emission rights trading remains immune to deliberate data modification and falsification, guaranteeing exceptionally high data stability and reliability. When two companies reach a deal in the carbon emission rights transaction, a new block node is created, which contains all the transaction data, such as the transaction object, transaction time, transaction amount and emission data. The data is verified by the transaction parties who sign the data before publishing it on the blockchain network. This process improves accountability and reliability of the data, thereby strengthening data security.

(iv) Decentralised transactions

Blockchain's decentralised attributes remove the need for intermediaries or middlemen in the trading process, reducing transaction costs and streamlining trading processes. A blockchain-enabled carbon trading system operates as an open, decentralised network ledger, distributing the storage and processing of data related to carbon trading across various network nodes, each with equal rights and obligations (Lin, 2016). Any changes in information, such as transactions, can be added by anyone using the same technical standards for bookkeeping, thus extending the blockchain network. Unlike the traditional 'star' topology, the decentralised mesh structure of blockchain technology resolves the issue of 'single point of failure' by replacing centralised hardware and significantly lowers the construction and operational costs of the carbon trading network market by removing third-party administrators (Ji, 2019). This decentralised system can achieve self-verification, transmission, and management of information, with data blocks in the blockchain being collectively maintained by distributed nodes to uphold them.

4. China's Exploration of Blockchain Applications for Carbon Trading

This section traces China's blockchain applications for carbon trading by three innovation spectrums, including the development of national/international standard plan, research and development (R&D) projects and early-stage pilot experiments. Figure 12.1 gives a panoramic view of China's ongoing blockchain applications for carbon trading initiatives from 2021.

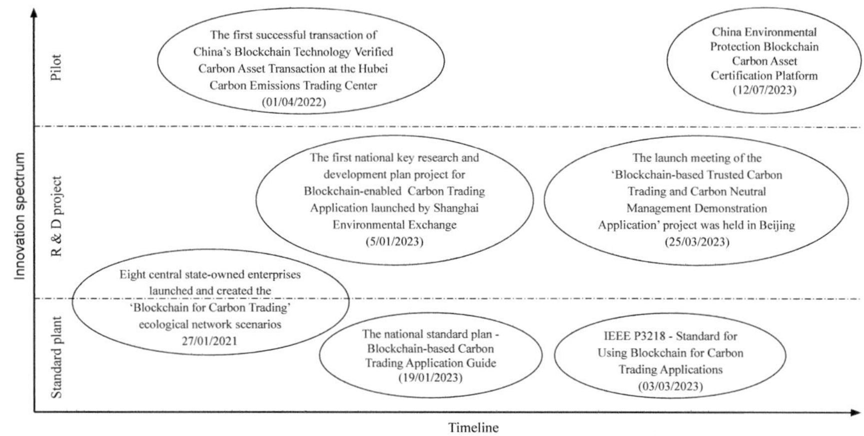

Fig. 12.1 China's ongoing initiatives for blockchain-enabled carbon trading. Diagram by the author based on multiple[1]

4.3 National/International Standard Plan

China has endeavoured to provide guidelines on the integration of blockchain technology in carbon trading by establishing state-wide and even international standards. This involves defining network protocols, interoperability guidelines, and technical frameworks to ensure a cohesive and interoperable trading ecosystem. On 27 January 2021, the first national standard plan – *Blockchain-based Carbon Trading Application Guide* – was released by eight central state-owned enterprises, including the State Grid and the National Energy Group. This guide identified several *Blockchain and Carbon Trading* ecological network scenarios that can inform the application of blockchain technology to carbon asset trading.

On 19 January 2023, another national standard plan – *Guide for Blockchain-based Carbon Trading Application* – was released under the jurisdiction of TC590 (National Technical Committee for Standardisation of Blockchain and Distributed

1 *Sources from:* http://www.tanpaifang.com/qukuailian/2023/030695084.html; https://std.samr.gov.cn/gb/search/gbDetailed?id=CB77487F21B39E37E05397BE0A0A47E6; https://journal.geidco.org.cn/cn/journalsDetailsCn/20230330/1641428840976879616.html; http://www.nengyuanjie.net/article/45620.html;https://www.cdmfund.org/30702.html; https://www.chaincatcher.com/article/2089967; and http://bj.news.cn/20230713/7c2686c0eadf43dea526c1ee8b1c287e/c.html.

Ledger Technology), governed by the Ministry of Industry and Information Technology. This standard specifies key processes of blockchain application in carbon trading services, such as registration on a blockchain network, identity authentication, registration and approval of carbon emission rights, and automatic matching and execution of carbon trading.

On 3 March 2023, China released the world's first *Blockchain and Carbon Trading* international standard, entitled *Standard for Using Blockchain for Carbon Trading Applications* (IEEE P3218). This standard was completed by a working group involving 17 IEEE member organisations, which summarises the technical requirements and principal specifications of blockchain in carbon trading application, relying on the blockchain public service platform – *Energy Carbon Chain* developed by State Grid Digital Technology Holding Company Limited. It helps to realise the traceability management of the life cycle of carbon trading in China and holds promise in addressing challenges such as the low efficiency of identity authentication for multiple entities in carbon trading and the complexities associated with data authentication.

4.4 Research and Development Project Initiatives

Several research and development (R&D) project initiatives have been conducted in China to explore the full potential of blockchain in carbon trading. These R&D project initiatives involve developing and testing blockchain-based solutions and frameworks to address challenges and enhance efficiency associated with carbon trading. On 5 January 2023, China's first national key R&D program project on blockchain-enabled carbon trading application was launched in Shanghai. This project is recognised as one of the highest-level R&D projects in China to date, consisting of five themes over the course of three years. It aimed to advance blockchain-based carbon trading applications in a real-world setting. On 28 March 2023, the discussion and implementation of the fourth theme – *Carbon Data Recording Devices and Accounting and Trading Mechanisms for Large-Scale Subscriber Access* was conducted by Yunnan Power Grid Corporation (YPG) and Shanghai Environmental Exchange (SEE). It provides an important research foundation and exploration experience for blockchain applications to facilitate carbon trading and carbon neutralisation, with Qujing, Chuxiong and Ruili in Dehong, Yunnan Province selected as three demonstration parks for applications. On 25 March 2023, the launch meeting of the *Blockchain-based Trusted Carbon Trading and Carbon Neutral Management Demonstration Application* project was organised by the Beijing Electric Power Trading Centre in Beijing. This project focuses on key technological innovation areas, such as regional carbon emissions identification and traceability, multi-subject low-cost access to authentication, and contractual and accurate accounting of carbon data. Four national industry parks in Beijing and Jiangsu were chosen for demonstration purposes, aimed at showcasing blockchain-based carbon trading applications in two key regions, i.e., Beijing-Tianjin Wing and Yangtze River Delta and creating replicable, scalable and demonstrable solutions and application benchmarks to other enterprises and the state.

4.5 Pilot Experiments

Pilot experiments involve implementing pilot programs and proof of concepts in real-world carbon trading scenarios to validate the potential of blockchain applications. The first successful transaction of China's Blockchain Technology Verified Carbon Asset Transaction was announced on 1 April 2022, at the Hubei Carbon Emissions Trading Centre. This transaction marks the world's first use of blockchain technology and Internet of Things (IoT) technology for carbon certification in a carbon transaction setting. This transaction not only spearheaded innovation in blockchain technology within the authentic carbon market but also translated theoretical designs into practical applications. On 12 July 2023, the China Environmental Protection Blockchain Carbon Asset Certification Platform designed and developed by China Environmental Protection Group was officially launched. It has become the first internet-based blockchain carbon asset management and verification platform in China, making an important contribution to the market-oriented blockchain-based carbon trading system. Although limited pilot use cases have been officially reported to date, it successfully validated the feasibility of applying blockchain technology to carbon market transactions through a tangible process, addressing practical challenges faced by the carbon trading market. This pilot success stands as a landmark for future endeavours, and the valuable learning experience gained can serve as a pivotal guide for the advancement of blockchain technology in carbon trading.

5. Challenges and Considerations

5.1 Challenges Associated with Blockchain Applications for Carbon Trading

While China's exploration of blockchain applications in carbon emission trading holds promise and presents advancements at multiple levels, there are several daunting challenges that need to be carefully considered before embracing blockchain technology in carbon trading. As a new technology that has not yet evolved into a mature one, at the early stage of the development of the blockchain-enabled carbon market, the upfront costs of applying blockchain technology to the carbon trading market may be high. There are also a couple of technological challenges, such as technical complexity, interoperability, and scalability, which are big hurdles to the development and implementation of blockchain technology for carbon trading. The decentralisation, security, and scalability of blockchain technology face a mutual constraint, creating what is known as the blockchain trilemma problem (Cao et al., 2022; Conti et al., 2019). Achieving all three objectives simultaneously is challenging due to the inherent characteristics of blockchain's distributed ledger system (Gao and Li, 2022). Pursuing decentralisation and security in blockchain-based carbon trading makes it difficult to attain scalability in the carbon trading market. The decentralised and secure features impose limitations on node computing and system capacity within the blockchain, hindering its ability to handle extensive data records associated with large-scale carbon trading. Conversely, prioritising scalability in the carbon market results in slow and challenging storage and computation of carbon transaction data when the block size grows rapidly and substantially (Lin and Liao, 2017). Furthermore, the nascent state of blockchain technology in the early stages could result in technical errors during transactions, causing confusion in the trading market. This may lead to transactions not being triggered promptly or, in some cases, incorrect transactions, ultimately jeopardising the interests of both parties involved in the carbon transaction. In addition to the challenges pertaining to technology itself, lack of adoption, especially the limitation of use cases identified in our study, signifies that there is limited successful story to build on. Furthermore, higher operation cost and limited technical competence are also barriers to blockchain adoption for carbon trading.

5.2 Considerations in Adopting Blockchain for Carbon Trading

Blockchain applications for carbon trading in China's context are dominantly driven by the government. The Chinese government has been making enormous efforts to showcase the viability of blockchain technology for carbon trading by actively issuing national and international standard plans, financially supporting research and development project initiatives and conducting pilot experiments. Although existing studies and projects demonstrate blockchain applications for carbon trading, these efforts are still at the early stages. Before the large-scale uptake of blockchain for carbon trading, it still needs to answer the question of whether blockchain technology is the right solution. Blockchain technology has its extensive application domains. However, it is not a silver bullet or a panacea for all industry applications. As far as many application scenarios in carbon trading are concerned, blockchain can only solve difficulties such as transparency, traceability, and integrity in the data transmission process, and it can do nothing to ensure authenticity of the data source. Blockchain possesses the attributes of data traceability and tamper resistance, ensuring the security and authenticity of data once stored in the blockchain. However, it itself lacks the capability to verify the authenticity of data before it is stored (Powell et al., 2021). The qualifications of carbon verification organisations in the market vary, potentially impacting the quality of an enterprise's carbon emission data. It should be noted that malicious tampering or miscalculation of carbon emission data by enterprises can result in the storage of erroneous data in the blockchain. Furthermore, Blockchain's inherent nature as a "trust machine" and its tamper-proof characteristics may contribute to the emergence of dataism and data solidification. This phenomenon, known as "automation bias," involves individuals blindly trusting information within the blockchain network while overlooking alternative sources of information (Han, 2021; Powell et al., 2022). Therefore, ensuring the authenticity of carbon trading data before it is uploaded to the blockchain poses a significant challenge for the advancement of blockchain carbon trading. This challenge may be solved by combining technologies such as the Internet of Things and AI with blockchain to solve verification issues such as multi-party data sharing, cross-verification, and confirmation.

Blockchain-based carbon trading requires a robust and inclusive regulatory framework to ensure adherence to compliance standards and prevent market manipulation. In its early developmental stage, existing regulatory policies and agencies' scope may not comprehensively cover the regulatory landscape, leaving room for potential violations like benefit transfers and data falsification. The decentralised nature of blockchain highlights the autonomy of network nodes, reducing external interference. However, this autonomy also amplifies the challenges associated with external regulation in the carbon emissions trading market. Therefore, it necessitates clear and comprehensive regulatory frameworks for blockchain-based carbon trading to ensure compliance and prevent market manipulation. This raises the question of whether China's existing highly centralised governance and legislation frameworks for managing the economy and information flows would suit a blockchain-based carbon trading landscape. If not, it is highly essential to identify and address the specific gaps and challenges that could hinder effective blockchain implementation in this sector.

Achieving a decent balance between transparency and data privacy needs to be particularly considered. Maintaining the right equilibrium is crucial to protect sensitive information, while facilitating information sharing in a more transparent manner. Blockchain's high transparency attributes ensure openness in the carbon trading market but poses a risk of privacy breaches. Excessive exposure of private data may expose trading partners to malicious entities, leading to potential harm to their economic assets and information security incidents. Therefore, the trade-off between transparency and data privacy is underscored especially during the initial stages of establishing the blockchain carbon market (Dorri et al., 2017). If this trade-off is well-managed, it could balance the need for open, transparent transactions with the protection of sensitive information in carbon trading. By carefully striking the balance between transparency and data privacy, the blockchain-based carbon trading market can achieve the dual goals of accountability and confidentiality, which can further pave the way for its successful large-scale implementation.

Conclusions

China, the world's largest carbon market, has endeavoured to embrace blockchain technology in carbon emission trading transactions, which reflects its ambition to build a transparent, efficient, and secure market for carbon trading. As a new technology with innovative potential, blockchain can provide a safer, more efficient and economic market environment as well as a visible, credible and reliable regulatory environment for carbon trading. The integration of blockchain technology has the potential to enhance the effectiveness and credibility of China's carbon market, build a perfect carbon trading mechanism, and indirectly promote the upgrading of the industrial structure of key carbon emitting entities, contributing to its broader climate goals. While China has seen limited instances of blockchain carbon trading to date, the introduction of national and international standards, coupled with ongoing research and development initiatives by enterprises and scientific research units, is expected to drive the continued development of blockchain carbon trading platforms. Anticipated growth in use cases is on the horizon. Numerous use cases are anticipated to emerge with the availability of national and international standards and the progress of research and development projects.

China's blockchain for carbon trading initiatives cover top-level standard plans, national R&D projects, and early-state pilot experiments. Looking further afield and into the future, the mix of diverse initiatives has paved the way for more blockchain applications for carbon trading to merge. The national and international standard plans, e.g., *Blockchain-based Carbon Trading Application Guide and Standard for Using Blockchain for Carbon Trading Applications* (IEEE P3218), pre-define the guidelines for the development of standardised solutions tailored to blockchain applications for carbon trading. The establishment of national and international standards provides a basis for China to build a carbon trading management system, solve many problems in the application of carbon trading, and improve blockchain technology and related specifications for the entire carbon trading industry. This can further ensure the interoperability of different blockchain applications for carbon trading and help address the "isolated island" problem for broad adoption of blockchain-based carbon trading solutions. This study is limited to the descriptive report on current developments of blockchain initiatives for carbon trading in China. However, we posit that these explorations, if successful, will strengthen China's ambition to actively carry out research on blockchain, carbon trading and carbon neutral management, prompt all-round and deep-level integration research, and promote the application requirements and scenario construction of blockchain to support credible carbon trading. Despite still being in the early-stage, explorations and pilot experiments can form a batch of scientific and technological achievements, which can serve as a model for other initiatives. These pilots can be also scaled up to be a more mature solution that has the potential for next-step broad adoption. The success of pilot experiments can be leveraged to propose a blockchain architecture that supports credible carbon transactions of multiple subjects and develops a credible carbon transaction and carbon neutral management platform based on blockchain.

China's blockchain infrastructure development and the ambitious action in adopting blockchain infrastructure for businesses and governments can be regarded as crucial drivers for blockchain-based carbon trading. For example, the blockchain-based service network – a general purpose infrastructure developed by China (Cao et al., 2020) – has been applied to cross-border trade (Cao et al.,

2022). Another example is that the government of Shanghai has implemented a strategic plan to develop its urban blockchain digital infrastructure system between 2023 and 2025.[3] These advancements highlight the established foundation for future explorations and developments, which include but are not limited to decentralised carbon marketplaces and price prediction and service discovery. Developing a decentralised carbon trading market has the potential to replace additional centralised hardware facilities or third-party management institutions and distributing the storage and processing of carbon trading-related data in different network nodes (Ji et al., 2021), providing a fairer and more efficient trading platform for carbon trading buyers and sellers. In addition to improving efficiency, transparency, and trust within the carbon trading ecosystem, blockchain technology holds the promise of facilitating a price signalling and service discovery mechanism that improves trading efficiency and incentivises emission reduction, thereby making the trading price more reasonable and accurate (Liu, 2020).

References

Blaufelder, C., Levy, C., Mannion, P. and Pinner, D. (2021, January 29). *A blueprint for scaling voluntary carbon markets to meet the climate challenge*. https://www.mckinsey.com/capabilities/sustainability/our-insights/a-blueprint-for-scaling-voluntary-carbon-markets-to-meet-the-climate-challenge

Cao, Boyen, Deane and Miller (2022). Enabling cross border trade in the face of regulatory barriers to data flow–the case of the Blockchain-based Service Network. *Blockchain for Industry*. https://eprints.qut.edu.au/229873/

Cao, M. (2020). A case study on China's carbon emission trading system: experiences and recommendations. *Asia Pacific Journal of Environmental Law*, *23*(2), 106–135.

Cao, S., Deane, F., Foth, M., Powell, W., Robb, L. and Turner-Morris, C. (2020). China's Blockchain Services Network–A new global emerging infrastructure. *ACS Informationage*. https://ia.acs.org.au/article/2020/blockchain-services-network.html

Cao, S., Foth, M., Powell, W., Miller, T. and Li, M. (2022). A blockchain-based multisignature approach for supply chain governance: A use case from the Australian beef industry. *Blockchain: Research and Applications*, 3(4), 100091. https://doi.org/10.1016/j.bcra.2022.100091.

Cao, S., Xu, H. and Bryceson, K. P. (2023). Blockchain Traceability for Sustainability Communication in Food Supply Chains: An Architectural Framework, Design Pathway and Considerations. *Sustainability: Science Practice and Policy*, *15*(18), 13486.

Chen, B. (2017). Lun Woguo Tanshichang Jinronghua de Jianguan Kunjing ji qi Zhidu Wanshan [On the Regulatory Dilemma of Financialization of China's Carbon Market and Its Institutional Improvement]. *Zhengquan Shichang Daobao*, *9*, 69–78.

Chen, J., Chen, S., Liu, Q. and Shen, M. I. (2021). Applying Blockchain Technology to Reshape the Service Models of Supply Chain Finance for SMES in China. *The Singapore Economic Review*, 1–18.

Conti, M., Gangwal, A. and Todero, M. (2019). Blockchain Trilemma Solver Algorand has Dilemma over Undecidable Messages. *Proceedings of the 14th International Conference on Availability, Reliability and Security*, Article 16.

[3] https://cointelegraph.com/news/shanghai-plans-for-blockchain-digital-infrastructure

Crawley, J. (2021, June 7). *China's Ministry of Industry and Information Technology Outlines Proposals for Blockchain Development*. https://www.coindesk.com/markets/2021/06/07/chinas-ministry-of-industry-and-information-technology-outlines-proposals-for-blockchain-development/

Cui, X. M., Li, X. H. and Tang, Y. J. (2016). Shehui Yali, Tanxinxi Pilu Toumingdu yu Quanyi Ziben Chengben [Social Pressure, Carbon Disclosure Transparency, and the Cost of Equity Capital]. *Dangdai Caijing*, *11*, 117–29.

de Larragán, J. de C. (2010). From the EU ETS to a Global Carbon Market: An Analysis and Suggestions for the Way Forward. *European Energy and Environmental Law Review*, *19*(1). https://kluwerlawonline.com/journalarticle/European+Energy+and+Environmental+Law+Review/19.1/EELR2010001

Dorri, A., Steger, M., Kanhere, S.S. and Jurdak, R. (2017). BlockChain: A Distributed Solution to Automotive Security and Privacy. *IEEE Communications Magazine*, *55*(12), 119–125.

Gao, Q.Q. and Li, Y. (2022). Lingtan Qukuailian: Luoji, Jiazhi yu Jinlu [Zero-carbon blockchain: logic, value and way forward]. *Tansuo Yu Zhengmeng*, *38*(1), 147–157+180.

Han, X.Z. (2021). Sifa Qukuailian de Fuhe Fengxian yu Shuangceng Guizhi [Compound Risks and Two-tier Regulation of Judicial Blockchain]. *Xian Jiaotong Daxue Xuebao(shehui Kexue Ban)*, *41*(1), 136–44.

Jiang, T., Song, J. and Yu, Y. (2022). The influencing factors of carbon trading companies applying blockchain technology: evidence from eight carbon trading pilots in China. *Environmental Science and Pollution Research International*, *29*(19), 28624–28636.

Ji, B., Chang, L., Chen, Z.H., Liu, Y., Zhu, D.H. and Zhu, L.Y. (2021). Jiyu Qukuailian Jishu de Dianli Tanpaifangquan Jiaoyi Shichang Jizhi Sheji yu Yingyong [Design and Application of Electricity Carbon Emission Right Trading Market Mechanism Based on Blockchain Technology]. *Dianli Xitong Zidonghua* , *45*(12), 1–10.

Ji, X.Q. (2019). Jiyu Qukuailian Jishu de Tanjinrong Shichang Fazhan Moshi Chutan [An Initial Exploration of Carbon Financial Market Development Model Based on Blockchain Technology]. *Jiazhi Gongcheng*, *38*(7), 193–96.

Lin, I.-C. and Liao, T.-C. (2017). A survey of blockchain security issues and challenges. *International Journal of Network Security*, *19*(5), 653. 659.https://doi.org/10.6633/IJNS.201709.19(5).01

Lin, X.X. (2016). Qukuailian Jishu zai Jinrongye de Yingyong [Blockchain technology in the financial industry]. *Zhongguo Jinrong*, *8*, 17–18.

Liu, T. (2020). Qukuailian Jishu dui Tanjiaoyi Jiage Jizhi de Yingxiang Yanjiu [Research on the impact of blockchain technology on carbon trading price mechanism]. *Jiage Lilun Yu Shijian*, *8*, 54–57.

Ma, G.S. and Duan, M.S. (2024). Huanjing Quanyi Jiaoyi zhong de Tanjianpai Xiaoying Chongfu Jisuan Fengxian ji Yingdui [The Risk of Double Counting of Carbon Emission Reduction Benefits in Environmental Equity Trading and Response to it]. *Qihou Bianhua Yanjiu Zhan*, *20*(1), 85–100.

Powell, W., Cao, S., Foth, M., He, S., Turner-Morris, C. and Li, M. (2022). Revisiting trust in supply chains: How does blockchain redefine trust? In B. Abdelaziz, I. Khalil, & B. Aouni (Eds.), *Blockchain Driven Supply Chains and Enterprise Information Systems*. Springer Nature.

Powell, W., Foth, M., Cao, S., & Natanelov, V. (2022). Garbage in garbage out: The precarious link between IoT and blockchain in food supply chains. *Journal of Industrial Information Integration*, 25, 100261. https://doi.org/10.1016/j.jii.2021.100261

Qin, M., Su, C.-W., Lobonţ, O.-R. and Umar, M. (2023). Blockchain: A carbon-neutral facilitator or an environmental destroyer? *International Review of Economics & Finance*, *86*, 604–15.

Tang, J.H., Wu, W.T. and Chen, X. (2023). Qukuailian Jishu zai Tanjiaoyi Tixi zhong de Yingyong Yanjiu [Research on the Application of Blockchain Technology in Carbon Trading System]. *Zhongguo Huanjing Guanli*, *15*(5), 78–86.

Tan, L. (2022, February 17). *The first year of China's national carbon market, reviewed*. https://chinadialogue.net/en/climate/the-first-year-of-chinas-national-carbon-market-reviewed/

Thomas, S., Dargusch, P. and Griffiths, A. (2011). The drivers and outcomes of the clean development mechanism in China. *Environmental Policy and Governance*, *21*(4), 223–39.

Wei, Y.P. and Yang, M. (2017). *Qiye Tanxinxi Pilu: Xianzhuang, Wenti ji Duice—Jiyu 2015nian Shenshi Shangshi Gongsi Nianbao de Tongji Fenxi [Corporate Carbon Disclosure: Current Situation, Problems and Countermeasures–Statistical Analysis Based on the Annual Reports of Listed Companies in Shenzhen City in 2015]. 10*, 110–114+4.

Wen, H.-X., Chen, Z.-R. and Nie, P.-Y. (2021). Environmental and economic performance of China's ETS pilots: New evidence from an expanded synthetic control method. *Energy Reports*, *7*, 2999–3010.

Yang, X., Zhang, J., Bi, L. and Jiang, Y. (2023). Does China's Carbon Trading Pilot Policy Reduce Carbon Emissions? Empirical Analysis from 285 Cities. *International Journal of Environmental Research and Public Health*, *20*(5). https://doi.org/10.3390/ijerph20054421

Yan, Z.Y. and Li, J. (2020). Jiyu Qukuailian Jishu de Tanpaifang Jiaoyi ji Jiankong Jizhi Yanjiu [Research on Carbon Emission Trading and Monitoring Mechanism Based on Blockchain Technology]. *Qiye Jingji*, *39*(6), 31–37.

Ye, Q., Gao, C.Y. and Jiang, G.X. (2022). Dashuju Huanjing xia Woguo Weilai Qukuailian Tanshichang Tixi Sheji [Design of China's future blockchain carbon market system under big data environment]. *Guanli Shijie*, *38*(1), 229–49.

Yoro, K.O. and Daramola, M.O. (2020). Chapter 1 - CO2 emission sources, greenhouse gases, and the global warming effect. In M. R. Rahimpour, M. Farsi, & M.A. Makarem (Eds.), *Advances in Carbon Capture* (pp. 3–28). Woodhead Publishing.

Zeng, J.G., Zhou, Y.H. and You, L. S. (2021). Wulianwang + Qukuailian jishu-Tanpaifang Toumingdu yu Chanye Shengji Biange [Internet of Things + Blockchain Technology, Carbon Transparency and Industrial Upgrading Changes]. *Zhongguo Fazhan*, *21*(S1), 83–88.

Zhang, Z. (2015). Carbon emissions trading in China: the evolution from pilots to a nationwide scheme. *Climate Policy*, *15*(sup1), S104–S126.

Zheng, R., Zhang, W. and Gao, Z.H. (2022). Jiyu Qukuailian Jishu de Zhengfu Shuju Kaifang Gongxiang Pingtai Goujian [Research on the Construction and Operation Mechanism of Government Data Open Sharing Platform Based on Blockchain Technology]. *Qingbao Kexue*, *40*(5), 137–43.

Zhou, L., Zhang, S.P., Hou, F.M. and Zhang, L.P. (2020). Jiyu Qukuailian Jishu de Tanjiaoyi Moshi Goujian [Construction of carbon trading model based on blockchain technology]. *Zhongguo Shuitu Baochi Kexue*, *18*(3), 139–45.

Zhu, L.H., Gao, F., Shen, M., Li, D.Y., Zheng, B.K., Mao, H.L. and Wu, Z. (2017). Qukuailian Yinsi Baohu Yanjiu Zongshu [A Review of Blockchain Privacy Protection Research]. *Jisuanji Yanjiu Yu Kaifa*, *54*(10), 2170–85.

Epilogue

Good Isn't so Easy: Reflections on a Decade of Aspiration for Blockchain Applications

Warwick Powell[1*]

The question of 'the good' and its relationship to technology invariably provokes reflections not so much about technology *per se*, but about what it is that is thought to be 'the good' (and its 'other'), and how such normative presuppositions are interpolated through the related discourses of various associated communities. 'Blockchain for good' is, in some senses, a recent variation on these dynamics, though its explicit labelling is provocative—or evocative—in itself, suggesting a context that has been, at best, normatively ambivalent; and if not ambivalent, normatively 'tainted.' In what conditions does a notion of 'blockchain for good' become invoked? Does its emergence speak of a concern that blockchain wasn't contributing to 'the good'? Or that blockchain could be used for something other than 'good'? Does labelling something go towards understanding 'it,' or does it fall foul of Primo Levi's caution that the mere act of naming something is in fact dangerous, as doing so leads one to believe that all has been explained when, in fact, nothing of the sort is the case?

Blockchain's own birth, announced in Nakamoto's seminal bitcoin white paper, was anything but ambivalent in normative presuppositions. The genesis block was imprinted with a mark of the times, referencing the global financial crisis of 2008, foretelling of the possibilities, the dreams and, dare one say, evangelical aspirations of bitcoin as a way out of the morass. The quest for digital cash *sans* centralised authority had long featured on the agendas of the cypherpunks (Jarvis, 2022), the anarcho-libertarians for whom the internet had lost its liberational vigour. The internet had, for them, been enlisted and subsumed by the various apparatus of state, and needed a reboot; it had to be reclaimed. The dream of wresting control over the issuance and governance of 'money' was never far away from this anti-dirigiste techno-movement.

The libertarian stripe was embedded in the foundational design and operational principles of blockchain *a la* Nakamoto. Not only did the bitcoin blockchain enable the tracking of transactions between parties without the intermediation of a centralised 'keeper of the ledger,' its operational solution enabled anyone, anywhere and at any time to participate in the maintenance of the ledger. No

[1] QUT Design Lab, Queensland University of Technology, Brisbane, Australia
[*] Corresponding author: w2.powell@qut.edu.au

permission was needed; no identity was required. Anonymous and permissionless were the *sine qua non* of a libertarian distributed ledger. No one was going to tell anyone else who could and couldn't play.

The distributed ledger also brought a range of attributes that, on face value, carried normative connotations. The ledger was censorship-resistant; pay the transaction fee and no-one could stop the update. No questions asked. The ledger was immutable; once updated, there was no going back to make alterations to the historic record. The mechanisms of updates, enacted by way of smart contracts, were also subject to these same conditions. Smart contracts were self-executing; they would, like vending machines, behave in predictable ways, over and over again. How much better could things get? No capricious control by way of asymmetric power relations; no risk of data malfeasance; human free transactions freed from the whims of human error, or worse, bad humans. A nirvana of truth and democratisation was only a few keystrokes away.

In this technical world, anchored by the mathematics of cryptography, trust was at best a five letter word. At worst, it was a dirty word. In this technical information context, the notion of trust denoted a situation in which the system's integrity was exposed to single points of failure. Nakamoto talked about being able to undertake payments without the need of a trusted intermediary; in other words, peer-to-peer. Yet, paradoxically, *trust* entered the blockchain community's lexicon through the backdoor of linguistic slippage, a Derrida-like trapdoor of *"differance:"* in a non-technical register, trust was seen as a good thing, and the problems of the world were—in part—a consequence of the absence of trust (Powell, 2022; Powell et al., 2023). In this parallel world, blockchain technology heralded an opportunity to reclaim and reinstall trust into social relations. Yet, the discussions about trust remain confused and superficial. The notion would simply be invoked as an under-theorised and taken-for-granted *good.*

So many *nice* words; words that could pull at the affective aspirations and wants of people all over the world. In the aftermath of the global financial crisis of 2008, there was a dearth of trust in the institutions of power that claimed to act in the interests of the multitudes. But not just a question of *trust.* The palpable sense of disenfranchisement; of asymmetry and powerlessness—remember *Occupy Wall Street*? (Van Gelder, 2011)—and systemic alienation called for a social and economic governance reboot. Along came Nakamoto. Just as his message of trustless-ness was garbled by people suffering from a loss of trust, so too was the idea of *consensus*. Like trust, *slippage* or an *indeterminacy* was part and parcel of the idea of consensus enabling a narrow technical concept to transform into a normative proposition. For Nakamoto, the consensus mechanism was a means by which the authenticity of a decentralised blockchain network could be maintained. Given the constraints of permissionless and anonymous participation, the risk of malfeasance behaviour could only be mitigated by means of *economic* incentives for doing the 'right thing' and disincentives for trying to do the 'wrong thing.' Block 'miners' would be rewarded, and as the chain got longer, the costs of winding back the clock became prohibitively costly. The mechanics of game theory were practically harnessed to keep things tight. Agent calculus, presumed to take the form of narrow cost-benefit rationalities, could be relied upon to drive the kinds of behaviours—the *good* behaviours, that is—that the blockchain ecosystem cherished. Out with Gordon Gecko's 'greed is good'[1] version of game theoretic behavioural parameters (remember, Wall Street's collapse presaged Nakamoto's arrival) and in with Nakamoto's version. *Homo economicus* proved to be a malleable subject.

An intellectual and design boom was catalysed, as game theoretic ideas permeated their way through the blockchain ecosystem body. It was as if everyone had discovered the Holy Grail of behavioural optimisation for normative objectives via financial incentives. *Cryptoeconomics* was born. Incentivising behaviours that a particular community believed was desirable was, no doubt, a useful way of framing an 'operating environment,' especially one for which a ledger of double-entry accounts backed by public-private key cryptography was at its beating heart. Market design and behavioural economics were already actively exploring the application of incentive mechanisms as ways of (1) seeking to understand observed behaviours that deviated from some theoretical notion of rational, and (2) design policies that could 'nudge' behaviours in aspirational directions. The blockchain ecosystem was living through a Zeitgeist shared by, and created with, others.

Carrots were relatively easy; *sticks*, on the other hand, were another thing altogether. Despite expectations of high security and promises of good behaviour, cases of hacking, theft and fraud soon became synonymous with the blockchain world. The vulnerabilities of a nascent technology were brutally exposed, no more so than in the high profile case of "The DAO project" (Dhillon et al., 2017), which was hacked in June 2016—a year after the Ethereum mainnet went live—when some US$60 million worth of ETH was stolen. This event set off a clash of normative presuppositions, which resolved itself by way of a split in the Ethereum community. The sanctity of the ledger, its immutability, was set against the reality of daylight robbery. For those who favoured the former, the idea of a forced rewrite of the ledger, to unwind the theft, was simply unthinkable and unconscionable. Caveat Emptor, they cried. There was to be no 'consensus.' A brute force resolution was occasioned; the network was forked.

The normative niceties of the idea of consensus could not mask the clash of principles laid bare by the DAO case. Indeed, like much else that appears to come with a normative good baked in, the contextual limitations were to always trump. The invocation of the idea of consensus raised more questions than it actually answered. The centrality of the idea, originally in narrow technical contexts of some mechanism to enable synchronisation between a network of computer nodes, was to open a Pandora's Box of governance problems to which there were, and are, no easy answers. If anything, and perhaps this is a good thing, the blockchain

---

[1] https://www.slashfilm.com/1176683/gordon-gekkos-famous-greed-is-good-line-was-inspired-by-an-actual-wall-street-criminal/

community—broadly speaking—was not like a marooned and isolated Robinson Crusoe in confronting these questions and challenges.

Constrained by the (self imposed) conditions of permissionless-ness and anonymity, the *problematique* of consensus was compelled to confront ways of dealing with malfeasance. The DAO case was one, but prevention was preferred to 'cure.' What was to be called 'staking' became the preferred mechanism; but in a case of 'what is old is new again,' the idea of staking was actually the rearticulation of older ideas of performance bonds within a self-enforcing smart contracts operating environment. Bad behaviour would be punished by way of deductions from the staked funds; in Ethereum parlance, the bad actor would have their account 'slashed.' In an anonymised set up, the temptation to do bad was only tempered by the relative trade-off between the 'slash' and the upside of misdemeanour. In fact, not much else was at stake. At stake? Yes, what's at stake? Monetary penalty was actually the only real enforceable penalty in an anonymised and digital world. And that's never been enough to stop bad behaviour. The Wall Street firms at the heart of the global financial crisis were fined; but few would suggest that has resulted in a change of ways.

Is *anonymity* a universal and context-free *good*? Good question, it seems. Indeed, is *permissionless* a similarly universal and context-free *good*? If the idea is of having something at stake as a mitigant against malfeasance, then what more could be put at stake? What is something that is immeasurably more valuable than money? Is it one's reputation? Or perhaps one's ability to participate in some economic activity or another in the future? Context matters, after all.

Aspirations of democratising participation in the governance of the commons, via what became known euphemistically as Distributed Autonomous Organisations (DAOs), have extended the theme of consensus beyond the foundations of maintenance of the ledger itself, to how decision-making can be enabled utilising the discursive and technical capabilities of blockchain technology ecosystems. Tokenisation and self-executing smart contracts laid the promise of participatory decision-making. Vote with tokens. Satisfying the decision threshold would automate execution of the decision. No central authority needed here. Voilà. Good, right? Yet, rather than usher in a new world of participatory democracy, the issues at stake in what counts for 'democratic' governance were progressively laid bare. What were the decision-making thresholds that should apply to what questions? What constitutes a sufficient 'majority'? A majority of those eligible to vote? A majority of eligible tokens? Is it a case of one wallet one vote, one person one vote, or one token one vote? What about a majority of votes cast, rather than of votes eligible? Is it a simple majority or a super majority?

If DAOs did not usher in an 'off the shelf' democratic utopia, their emergence has reawakened interest in alternative forms of governance and organisational establishment. In contrast to stock companies, DAOs rekindle an interest in forms of social organisation that predate the stock company; membership based associations, guilds, mutuals and cooperatives look more like DAOs than do stock companies. So, what is old is new again. That's a good thing, if the stock company form and its incentive mechanisms made an oversized contribution to the financial crisis against which Nakamoto reacted.

Invocations of the good presupposes some notion of what constitutes good and what doesn't. That those in the blockchain ecosystem have consciously engaged in activities that open these questions up for reflection, in theory and in practice - as a form of *praxis*, so to speak - is a good thing. That it is a good thing is also a question of context, of history, of situation. That Nakamoto's intervention was a timely one, an intervention at a specific conjuncture, doubtless enabled the technology to be catapulted into a wider sphere of human imagination. The social and economic turmoil occasioned in the shadow of the global financial crisis provoked reactions across many parts of the world; the 'system', however it was understood, had failed to deliver on the promise. The palpable inequities that made Thomas Piketty's (2014) *Capital in the 21st Century* a bestseller provoked a political and affective reaction. Blockchain's promise to disintermediate was grasped as a way of sidestepping the powers that be, those that were in the eyes of many responsible for the failures of the system to date, and to deliver humanity to the promised land of peer-to-peer integrity and opportunity for all.

Yet, as the decade rolled on, from the time the Ethereum network's genesis block was minted in mid-2015, the answers to the riddles of real world system complexity, economic and social malfeasance and persistent organisational and governance dysfunctionality did not seem as easy as many of the early proselytisers would have had one believe. Whatever the *good* may be, it remains as elusive as ever.

References

Dhillon, V., Metcalf, D. and Hooper, M. (2017). The DAO Hacked. In V. Dhillon, D. Metcalf, & M. Hooper (Eds.), *Blockchain Enabled Applications: Understand the Blockchain Ecosystem and How to Make it Work for You* (pp. 67–78). Apress. https://doi.org/10.1007/978-1-4842-3081-7_6

Jarvis, C. (2022). Cypherpunk ideology: objectives, profiles, and influences (1992–1998). *Internet Histories*, *6*(3), 315–42. https://doi.org/10.1080/24701475.2021.1935547

Piketty, T. (2014). *Capital in the Twenty-First Century*. Harvard University Press.

Powell, W. (2022). *China, Trust and Digital Supply Chains: Dynamics of a Zero Trust World*. Routledge. https://doi.org/10.4324/9781003184614

Powell, W., Cao, S., Foth, M., He, S., Turner-Morris, C. and Li, M. (2023). Revisiting Trust in Supply Chains: How Does Blockchain Redefine Trust? In A. Bouras, I. Khalil, & B. Aouni (Eds.), *Blockchain Driven Supply Chains and Enterprise Information Systems* (pp. 21–42). Springer International Publishing. https://doi.org/10.1007/978-3-030-96154-1_2

Van Gelder, S. (2011). *This Changes Everything: Occupy Wall Street and the 99% Movement*. Berrett-Koehler Publishers.

Index

For Product Safety Concerns and Information please contact our
EU representative GPSR@taylorandfrancis.com Taylor & Francis
Verlag GmbH, Kaufingerstraße 24, 80331 München, Germany